AF775614

Springer Theses

Recognizing Outstanding Ph.D. Research

For further volumes:
http://www.springer.com/series/8790

Aims and Scope

The series "Springer Theses" brings together a selection of the very best Ph.D. theses from around the world and across the physical sciences. Nominated and endorsed by two recognized specialists, each published volume has been selected for its scientific excellence and the high impact of its contents for the pertinent field of research. For greater accessibility to non-specialists, the published versions include an extended introduction, as well as a foreword by the student's supervisor explaining the special relevance of the work for the field. As a whole, the series will provide a valuable resource both for newcomers to the research fields described, and for other scientists seeking detailed background information on special questions. Finally, it provides an accredited documentation of the valuable contributions made by today's younger generation of scientists.

Theses are accepted into the series by invited nomination only and must fulfill all of the following criteria

- They must be written in good English.
- The topic should fall within the confines of Chemistry, Physics, Earth Sciences, Engineering and related interdisciplinary fields such as Materials, Nanoscience, Chemical Engineering, Complex Systems and Biophysics.
- The work reported in the thesis must represent a significant scientific advance.
- If the thesis includes previously published material, permission to reproduce this must be gained from the respective copyright holder.
- They must have been examined and passed during the 12 months prior to nomination.
- Each thesis should include a foreword by the supervisor outlining the significance of its content.
- The theses should have a clearly defined structure including an introduction accessible to scientists not expert in that particular field.

Joseph T. Lizier

The Local Information Dynamics of Distributed Computation in Complex Systems

Doctoral Thesis accepted by
The University of Sydney, Australia

Springer

Author
Dr. Joseph T. Lizier
CSIRO ICT Centre
Marsfield, NSW
Australia

Supervisors
Dr. Mikhail Prokopenko
CSIRO ICT Centre
Marsfield, NSW
Australia

Prof. Albert Zomaya
School of Information Technologies
The University of Sydney
Sydney, NSW
Australia

ISSN 2190-5053 ISSN 2190-5061 (electronic)
ISBN 978-3-642-32951-7 ISBN 978-3-642-32952-4 (eBook)
DOI 10.1007/978-3-642-32952-4
Springer Heidelberg New York Dordrecht London

Library of Congress Control Number: 2012945730

© Springer-Verlag Berlin Heidelberg 2013
This work is subject to copyright. All rights are reserved by the Publisher, whether the whole or part of the material is concerned, specifically the rights of translation, reprinting, reuse of illustrations, recitation, broadcasting, reproduction on microfilms or in any other physical way, and transmission or information storage and retrieval, electronic adaptation, computer software, or by similar or dissimilar methodology now known or hereafter developed. Exempted from this legal reservation are brief excerpts in connection with reviews or scholarly analysis or material supplied specifically for the purpose of being entered and executed on a computer system, for exclusive use by the purchaser of the work. Duplication of this publication or parts thereof is permitted only under the provisions of the Copyright Law of the Publisher's location, in its current version, and permission for use must always be obtained from Springer. Permissions for use may be obtained through RightsLink at the Copyright Clearance Center. Violations are liable to prosecution under the respective Copyright Law.
The use of general descriptive names, registered names, trademarks, service marks, etc. in this publication does not imply, even in the absence of a specific statement, that such names are exempt from the relevant protective laws and regulations and therefore free for general use.
While the advice and information in this book are believed to be true and accurate at the date of publication, neither the authors nor the editors nor the publisher can accept any legal responsibility for any errors or omissions that may be made. The publisher makes no warranty, express or implied, with respect to the material contained herein.

Printed on acid-free paper

Springer is part of Springer Science+Business Media (www.springer.com)

Parts of this thesis have been published in the following articles:

J. T. Lizier, M. Prokopenko, and A. Y. Zomaya, "Information transfer by particles in cellular automata," in *Proceedings of the Third Australian Conference on Artificial Life, Gold Coast, Australia*, ser. Lecture Notes in Artificial Intelligence, M. Randall, H. A. Abbass, and J. Wiles, Eds., vol. 4828, pp. 49–60. Berlin/Heidelberg: Springer, 2007.

J. T. Lizier, M. Prokopenko, and A. Y. Zomaya, "Detecting non-trivial computation in complex dynamics," in *Proceedings of the 9th European Conference on Artificial Life (ECAL 2007), Lisbon, Portugal*, ser. Lecture Notes in Artificial Intelligence, F. Almeida e Costa, L. M. Rocha, E. Costa, I. Harvey, and A. Coutinho, Eds., vol. 4648, pp. 895–904. Berlin/Heidelberg: Springer, 2007.

J. T. Lizier, M. Prokopenko, and A. Y. Zomaya, "Local information transfer as a spatiotemporal filter for complex systems," *Phys. Rev. E*, vol. 77, no. 2, p. 026110, 2008.

J. T. Lizier, M. Prokopenko, and A. Y. Zomaya, "The information dynamics of phase transitions in random Boolean networks," in *Proceedings of the Eleventh International Conference on the Simulation and Synthesis of Living Systems (ALife XI), Winchester, UK*, S. Bullock, J. Noble, R. Watson, and M. A. Bedau, Eds., pp. 374–381. Cambridge, MA: MIT Press, 2008.

J. T. Lizier, M. Prokopenko, I. Tanev, and A. Y. Zomaya, "Emergence of glider-like structures in a modular robotic system," in *Proceedings of the Eleventh International Conference on the Simulation and Synthesis of Living Systems (ALife XI), Winchester, UK*, S. Bullock, J. Noble, R. Watson, and M. A. Bedau, Eds., pp. 366–373. Cambridge, MA: MIT Press, 2008.

J. T. Lizier, J.-D. Haynes, J. Heinzle, and M. Prokopenko, "Directed information structure in inter-regional cortical interactions in a visuomotor tracking task," *BMC Neuroscience*, vol. 10, no. Suppl 1, p. P117, 2009, *Proceedings of the Eighteenth Annual Computational Neuroscience Meeting Computational Neuroscience 2009 (CNS*2009)*, Berlin, Germany.

J. T. Lizier, M. Prokopenko, and D. J. Cornforth, "The information dynamics of cascading failures in energy networks," in *Proceedings of the European Conference on Complex Systems (ECCS), Warwick, UK*, p. 54, 2009, ISBN: 978-0-9554123-1-8.

J. T. Lizier and M. Prokopenko, "Differentiating information transfer and causal effect," *European Physical Journal B*, vol. 73, no. 4, pp. 605–615, 2010.

J. T. Lizier, M. Prokopenko, and A. Y. Zomaya, "Information modification and particle collisions in distributed computation," *Chaos*, vol. 20, no. 3, p. 037109, 2010.

J. T. Lizier, J. Heinzle, A. Horstmann, J.-D. Haynes, and M. Prokopenko, "Multivariate information-theoretic measures reveal directed information structure and task relevant changes in fMRI connectivity," *Journal of Computational Neuroscience*, vol. 30, no. 1, pp. 85–107, 2011.

J. T. Lizier, M. Prokopenko, and A. Y. Zomaya, "Coherent information structure in complex computation," *Theory in Biosciences*, vol. 131, no. 3, pp. 193–203, 2012.

J. T. Lizier, M. Prokopenko, and A. Y. Zomaya, "Local measures of information storage in complex distributed computation," *Information Sciences*, vol. 208, pp. 39–54, 2012.

For Amanda,
because I still like to impress you

Supervisor's Foreword

In early 1990s, Chris Langton created the term "computation at the edge of chaos". In doing so, he not only dissected computation into its three primitive functions: the transmission, storage, and modification of information, but also opened the way to study computation in the most generic form—information-theoretically. Moreover, a challenge was thrown out: how does one explain emergence of computation in a *dynamic* setting, and how is it related to *complexity* of the system in point?

It is a pleasure to introduce Dr. Joseph Lizier's thesis "The local information dynamics of distributed computation in complex systems". This work is not shy of taking on and meeting Langton's challenge: it goes to the heart of the phenomenon of computation, uncovers and analyses information-theoretic roots of each of the primitives, and brings these parts back together in a multiplicity of ways, revealing different shapes that coherence and complexity may take.

The dissertation leads us through complex, diverse, and dynamic domains that share an intricate fabric of computation: cellular automata (CA), random Boolean networks, brain's cortical networks, heart- and breath-rate interactions in sleep apnoea, modular robotic systems, cascading failures in power grids, etc. It reveals inner workings of these systems through their coherent use of memory ("active information storage"), communications ("information transfer"), and processing ("information modification").

Specifically, the *active information storage* is introduced, quantifying the information storage component that is directly in use in the computation of the next state of a process. The obtained profiles of local active information storage in CA provide evidence that blinkers and background domains are dominant information storage processes in these systems, and reveal that local storage may sometimes misinform the observer when the computation is dominated by some other primitives.

Significantly, the work introduces a measure of local *information transfer*, applied to CA for filtering coherent structures—the measure provides the first quantitative evidence for the long-held conjecture that particles (both gliders and domain walls) are the dominant information transfer agents in CA. This is further developed into a novel method for interregional brain connectivity analysis, using multivariate extensions to the mutual information and transfer entropy. This method is distinguished in using asymmetric, multivariate, information-theoretical analysis, which captures not only directional and non-linear relationships, but also collective interactions.

Bringing storage and transfer together in state-space diagrams, this dissertation distinguishes complex computation via *coherent structure* in local information dynamics profiles, and identifies both clear and "hidden" coherent structure via the state-space diagrams of the local information dynamics.

Building up on these measures, the study investigates the well-known *phase transition* between ordered and chaotic behavior in Random Boolean Networks, demonstrating maximisations in information storage and information transfer on either side of the critical point—thus, explaining the phase transition in terms of the intrinsic distributed computation.

Touching on the subject of *causality*, the work contrasts transfer entropy and information flow, which can be used separately to quantify information transfer and causal information flow respectively, and shows that causal information flow is a primary tool to describe the causal structure of a system, while information transfer can be used to describe the emergent computation in the system.

Subsequently, *information modification* is quantified at each spatiotemporal point in a system—via separable information, a novel measure which locally identifies events where separate inspection of the sources to a computation is misleading about its outcome (as the points where "the whole is greater than the sum of the parts").

In summary, the work presented in this dissertation forms a new sub-field—"information dynamics". It enables a rigorous characterization of diverse computation processes on a local spatiotemporal scale within a system, and the results provide important insights into the fundamental nature of distributed computation and the dynamics of complex systems.

After reading this thesis, no longer may one entertain a singular view on complexity or hope to measure the latter with a single static measure—instead one needs to consider information dynamics within a rich multi-faceted space, where each primitive (storage, transfer, and modification) becomes a separate dimension. Thus, complexity acquires shapes and trajectories defined by these axes, while coherence can be quantitatively studied via suitably defined state-diagrams.

I have no doubt that this study is at the forefront of the fundamental research that will eventually lead to a comprehensive information- and computation-theoretic framework unifying multiple complex physical, biological, social, and technological systems.

Sydney, 7 July 2012 Mikhail Prokopenko

Acknowledgments

A wise friend recently told me: “You never finish a Ph.D. thesis, you just decide to stop.” Despite the endless frustrations that make research what it is, I have had such a wonderful time during my Ph.D. studies that “deciding” to stop was rather difficult. As such, I would like to thank those who made this work possible and so enjoyable.

To my supervisor Mikhail Prokopenko at Australia’s Commonwealth Scientific and Industrial Research Organisation (CSIRO), I thank you first for the web page of the “Entropy discussion group” that I stumbled across and met you through. For the original concept that information transfer in complex systems was a hot topic worth pursuing, and that we should harness it in guided self-organization. For a countless number of interesting and valuable discussions, and your patience with my decisions (“she’ll be right, mate”). For your wise and diplomatic strategies, and support in finding funding for my ventures. For insightful feedback on my papers and these chapters, and mentoring on how to be a scientist. For balancing supervision and mateship. The debt I owe you is very large indeed.

To my academic supervisor Albert Zomaya at The University of Sydney (USyd), I thank you for always finding room for me in a very hectic schedule. I particularly appreciate your effort in maintaining our “long-distance relationship” for much of the time. Thank you for your insightful suggestions, and creative perspectives on new horizons for us to exploit. For your strong support and trust in my judgement, and the productive environment of your group. For the motivating and thoroughly enjoyable consultations, and the many laughs. I am grateful for all that I have learned from you.

In addition to my supervisors, I thank the others with whom I have collaborated on research during these years, much of which does not appear in this thesis: Ivan Tanev, Larry Yaeger, Mahendra Piraveenan, Dany Pradhana, Oliver Obst, X. Rosalind Wang, Peter Wang, Vadim Gerasimov, Jakob Heinzle, John-Dylan Haynes, Annette Horstmann, Valerio Sperati, Jürgen Pahle, John Mahoney, Olaf Bochmann, Gregor Obernosterer, Joseph DeRosa, Ben Mazzotta, Luciano Oviedo, Alexander Shpunt, Carver Tate, David Cornforth, Joschka Bödecker, N. Michael Mayer, and Mikhail Burtsev. I am grateful for the many discussions, including helpful comments and suggestions on my manuscripts and clarifications of their own work, from my collaborators above as well as from the following: Daniel Polani, Nihat Ay, Ralf Der, Melanie Mitchell, Thomas Schreiber, David Feldman, Cosma Shalizi, Kristian Lindgren, Andre Ribeiro, Alexander Kraskov, Paolo Crucitti, Masatoshi Funabashi, Christian Darabos, Alex Healing, Mikail Rubinov, and Antonio Lafusa. Similarly, I also particularly thank: Selvakennedy Selvadurai for being an additional associate supervisor in the early days of my candidature; office mates Mahendra Piraveenan, Astrid Zeman, and Khaled Almi'Ani; lunch-mates Oliver Obst and Rosalind Wang; members of CSIRO's Entropy discussion group; the Advanced Network Research Group (ANRG), and Distributed Computing discussion group at USyd; my early scientific mentors Ferg Brand and Graham Town; Terry Dawson and Mary Cudmore for helping put the title "complex systems science" to my interest in this area; the anonymous reviewers whose comments have improved the quality of my work, but in particular Daniel Polani, Ralf Der, and David Feldman for the incredibly useful (formal) reviews they provided for this thesis; Leontina Di Cecco at Springer for her help in publishing this thesis, and Michael Wibral for the suggestion and encouragement to do so.

Acknowledged contributions of others here include (specific references are included at the relevant chapters): Mirek Wójtowicz's *Java cellebration* for simulating and plotting cellular automata (e.g. in Fig. 3.4); Carlos Gershenson's *RBNLab* for simulating random Boolean networks; Ivan Tanev's original *snakebot* and genetic programming implementation; Jakob Heinzle, John-Dylan Haynes and Annette Horstmann from the Bernstein Center for Computational Neuroscience in Berlin provided the fMRI data set analysed in Sect. 8.2 (these authors conceived and performed the original experiment to obtain the data, obtained informed written consent from the subjects, and pre-processed the data as described in Sect. 8.2.3.1); Jean-Baptiste Rouquier's *CimulA* for analysing statistical complexity in CAs and plotting Fig. 2.2; *Cytoscape* for generating Figs. 8.3 and 8.4.

For financial support, I primarily thank: The University of Sydney for the award of the A.E. and F.A.Q. Stephens scholarship, and the Stephens family for the bequests that made this possible; and CSIRO for their Postgraduate Award, as a top-up scholarship. This financial support enabled me to study full-time. I would like to thank various sources of travel funding, the use of which has had a significant impact on the maturity of my work: CSIRO for funding via the Postgraduate Award, the Centre for Complex Systems Science, and the Energy Transformed flagship; the USyd under the Postgraduate Research Support Scheme (PRSS); the Australian Research Council's Complex Open Systems Network (COSNet); the Santa Fe Institute through NSF Grant No. 0200500 entitled "A Broad Research Program in the Sciences of Complexity"; and the organizers of ALifeXI for a travel bursary. In terms of resources, I thank CSIRO's Information and Communication Technologies (ICT) Centre for the laptop and other workspace arrangements in hosting me, and CSIRO's Advanced Scientific Computing group for access to high performance computing resources (the burnet cluster) used for simulation and analysis in this work. I also thank the staff of School of Information Technologies at USyd for facilitating my studies.

On the personal front, I thank my parents Pamela and George for the upbringing, education, sacrifices, and, along with my sister Sally, for the environment and encouragement that made me the person who could pull this off.

And of course my enormous gratitude goes to my wife and inspiration Amanda. First for the encouragement in continually asking me when I was going to enrol in my Ph.D. The first time I truly believed I would do it was when you started telling other people that was the plan, and that changed everything. For your support and stability, both emotional and financial. For your patience, and for more hours of proof-reading about this information thing than you deserved. Most of all for sharing these fabulous years with me. I would not trade a minute of it. I only hope that I can make as much of a contribution to your Ph.D. studies as you have made to mine.

Contents

Symbols

Typefaces

X, Y, Z	Variable names
x, y, z	Specific values taken by the variables X, Y, Z
$\hat{x}, \hat{y}, \hat{z}$	Values imposed on the variables X, Y, Z
$\mathbf{X}, \mathbf{Y}, \mathbf{Z}$	Sets of variables
$\mathbf{x}, \mathbf{y}, \mathbf{z}$	Values to the sets $\mathbf{X}, \mathbf{Y}, \mathbf{Z}$

Notation

x_n	Value of variable X at time n
$x_{i,n}$	Value of variable X_i at time n in a spatially-extended system $\mathbf{X}$
$\mathbf{x}_n$	Joint value of a spatially extended system $\mathbf{X}$ at time n
$x_n^{(k)}$	Joint value of the *past* k values of variable x, up to and including time n
$x_n^{(k+)}$	Joint value of the *future* k values of variable x, from time n onwards
$p(a)$	Probability of the event a
$p(a, b)$	Probability of the joint event $a \wedge b$
$p(\mathbf{a})$	Probability of the joint event $\mathbf{a}$
$p(a \mid b)$	Probability of the event a *given* event b
$\hat{p}_r(a)$	Probability of the *continuous* value a, *estimated* using parameter r
H_X	An example *average* information-theoretic measure on variable X
$h_X(n, ...)$	An example *local* information-theoretic measure on variable X at time n (lower-case by convention)
$h(x_n, ...)$	Alternate notation for a *local* information-theoretic measure on variable X at time n (lower-case by convention)
$h(i, n, ...)$	An example *local* information-theoretic measure in a spatially-extended system, on variable x_i at time n (lower-case by convention)

$\langle x_n \rangle_n$	Expectation value of x_n, averaged over all n
$\mathbf{V}_X$	Causal parents of variable X
$\mathbf{V}_X^Y$	Causal parents of variable X except for variable Y and X itself

Information-Theoretic Measures

The following table indicates the notation for several *average* information-theoretical measures used (both existing and introduced measures). The notation for their *local* values takes the lower-case of the same variable name, as indicated in the list of notation above.

H_X	Shannon entropy of variable X
$H_{X,Y}$	Joint entropy of X and Y
$H_{X\|Y}$	Conditional entropy of X given Y
$I_{X;Y}$	Mutual information of X and Y
$I_{\mathbf{X}}$	Multi-information of $\mathbf{X}$
$I_{X;Y\|Z}$	Conditional mutual information of X and Y given Z
$H_{\mu X}$	Entropy rate of X
E_X	Excess entropy of X
A_X	Active information storage of X
$T_{Y \to X}$	(Apparent) Transfer entropy from Y to X
$T_{Y \to X\|Z}$	Conditional transfer entropy from Y to X given Z
$T^c_{Y \to X}$	Complete transfer entropy from Y to X
T_X	Collective transfer entropy to X
$I_p(Y \to X \mid \hat{\mathbf{S}})$	Information flow from Y to X imposing $\mathbf{S}$
U_X	Intrinsic uncertainty of X
S_X	Separable information of X
D_X	Information destruction at X
I_X^{ss}	Multi-information in the state-space of X

Abbreviations

1D	One-dimensional
ALife	Artificial Life
BOLD	Blood Oxygen Level-Dependent (time-series)
CA	Cellular Automata
CRBN	Classical Random Boolean Network
ECA	Elementary Cellular Automata
EEG	Electroencephalogram
fMRI	Functional Magnetic Resonance Imaging
GRN	Gene Regulatory Network
MI	Mutual Information

PDF	Probability Distribution Function
RBN	Random Boolean Network
RTZ	Return To Zero (bit operation)
TE	Transfer Entropy
XOR	Exclusive OR (bit operation)

Chapter 1
Introduction

> "Although they (complex adaptive systems) differ widely in their physical attributes, they resemble one another in the way they handle information. That common feature is perhaps the best starting point for exploring how they operate." Gell-Mann, 1994 [1].[1]

The nature of **distributed computation** has long been a topic of interest in complex systems science, physics, artificial life, bio- and neuroinformatics. In all of these relevant fields, distributed computation is generally discussed in terms of **memory**, **communication**, and **processing**:

- **Memory** refers to the storage of information by an agent or process to be used in its future. It has been investigated in coordinated motion in modular robots [2], in the dynamics of inter-event distribution times [3], and in synchronisation between coupled systems [4].
- **Communication** refers to the transfer of information between one agent or process and another. It has been shown to be of relevance to biological systems (e.g. dipole-dipole interaction in microtubules [5], and in signal transduction by calcium ions [6]), social animals (e.g. schooling behaviour in fish [7]), and agent-based systems (e.g. the influence of agents over their environments [8], and in inducing emergent neural structure [9]).
- **Processing** refers to the combination or modification of stored and/or transmitted information into a new form. It has been discussed in particular for biological neural networks and models thereof [10–13] (where it has been suggested as a potential biological driver), and also regarding collision-based computing (e.g. [14, 15], and including soliton dynamics and collisions [16]).

Distributed computation is *any process conducted by multiple agents or entities that involves these* **operations on information**. Notable examples include: the time evolution of discrete dynamical systems such as cellular automata [17], information processing in the brain [18], gene regulatory networks computing cell behaviours [19], flocks or schools computing their collective heading [7], ant colonies computing the most efficient routes to food sources [20], or collective behaviour in artificial self-organised systems [2].

---

[1] With the use of "they", Gell-Mann was referring to complex adaptive systems; the term was inserted by this author in parentheses here for clarity.

J. T. Lizier, *The Local Information Dynamics of Distributed Computation in Complex Systems*, Springer Theses, DOI: 10.1007/978-3-642-32952-4_1,

© Springer-Verlag Berlin Heidelberg 2013

Indeed, these operations exist in all systems, whether or not the system was explicitly designed to compute or appears to be performing any useful function: this is referred to as *intrinsic computation* [21, 22]. It is why the universe can be seen to be computing its own future, as per Lloyd [23]: "What does the universe compute? It computes itself." This understanding of intrinsic computation also underpins why "information is physical and physics is information" [24].

Significantly, these three operations on information are formally primitive functions of Turing universal computation [25]:

- **information storage**,
- **information transfer** (or transmission), and
- **information modification**.

These operations are particularly important from a *theoretical perspective* in **complex systems science**, where they are the subject of a number of important conjectures regarding the fundamental nature of distributed computation and its relationship to emergent complex behaviour. One focus of such discussion is the **dynamics of computation**, i.e. the manner in which computations *unfold in time* and are *distributed across space*. This focus considers computation in cellular automata (CAs) [26, 27] in particular, and the manner in which information manipulation is said to be facilitated by interaction of emergent coherent structures known as particles. Conjecture holds that stationary particles (blinkers) implement information storage, moving particles (gliders and domain walls) implement information transfer, and particle collisions implement information modification. These conjectures are used to interpret computation not only in CAs, but in other systems where analogues of these coherent structures interact (e.g. in the brain [18]). Also, emergent complex behaviour has been postulated to be associated with the capability to support universal computation [25, 28, 29], with maximum capacity for these primitive functions said to occur near order-chaos phase transitions [10, 30]. This is generally referred to as the *edge of chaos* hypothesis [25, 31].

Yet despite the obvious theoretical and practical importance of these primitive operations on information, we have no framework for either quantifying them individually, quantifying their dynamics more specifically, or understanding how they interact to give rise to distributed computation. It is particularly noteworthy that this is also despite significant interest in and an abundance of general measures of complexity of computation in the literature (e.g. [32–36]).

The lack of such a framework is a serious impediment to our ability to understand distributed computation and the nature of complex behaviour. On a practical level, this limits the extent to which complex systems science can provide insights into the dynamics of computation in the systems described above, e.g. in measuring when and where information is transferred in the brain during cognitive tasks. It also limits the extent to which complex systems science can provide insights about computation that can be compared across systems. Furthermore, this leaves us without the required tools to design desired distributed computation in artificial self-organised systems, which is known to be a difficult task [37].

1.1 Hypothesis and Objectives

The **hypothesis** of this thesis is that *if we can describe and quantify distributed computation in terms of information storage, transfer and modification, then we will be better able to understand distributed computation in nature and its sources of complexity*.

Our primary objective then is to *define a complete framework* that quantifies the fundamental operations of information storage, transfer and modification in distributed computation. We will establish proper definitions for each, and clarify their relationships with similar concepts (e.g. how information transfer is distinct from causal effect). In particular, we seek to quantify these operations on a *local scale* in space and time in order to describe the *dynamics* of computation. We will refer to this as a **framework for the information dynamics of distributed computation**. We introduce the term **information dynamics** as "the study of operations on information by agents in distributed computation, and the dynamics of these operations on a local scale in space and time".

The framework will be used to provide insights into the *fundamental nature of distributed computation*, and how information is manipulated in complex systems. We will apply the framework to sample theoretical systems, such as cellular automata. We expect the framework to provide quantitative evidence for the well-known conjectures regarding the role of emergent coherent structures in computation in CAs. We will also seek to describe how the information dynamics interrelate to produce distributed computation, and the conditions under which such computation can be described as complex. For example, we will investigate whether these operations are indeed maximised in order-chaos phase transitions.

We will apply the framework to systems of practical interest, in particular natural systems, in order to explore our assertion that it can **answer meaningful questions** about the computation they are undertaking. In particular, we will explore the assertion that *local* information dynamics can provide more detailed insights than averaged measures. A typical question where local information dynamics could provide new insights is "*when* and *how much* information is transferred between two brain regions?"

Finally, we will attempt to use the framework to guide the **design** of artificial self-organised systems. We will explore the assertion that viewing the goal of the system as a distributed computation will allow significant new insights to be gained from our framework.

1.2 Approach

Our approach is based in **complex systems science**: the study of large collections of (generally simple) entities, where the global behaviour is a non-trivial result of the local interactions of the individual elements. Viewing the global behaviour of complex systems as being the result of a distributed computation between the individual elements is becoming increasingly popular (as per the opening quote from Gell-Mann [1] and e.g. [25, 26, 38, 39]).

Cellular automata have been an important focus for this perspective as model systems offering a range of dynamical behaviour, including supporting complex computations and the ability to model complex systems in nature [26]. Importantly from our perspective, there is very clear *qualitative* identification of emergent structures representing information storage, transfer and modification therein (e.g. [25, 26]). CAs are a critical proving ground for any theory on the nature of distributed computation: significantly, Von Neumann was known to be a strong believer that "a general theory of computation in 'complex networks of automata' such as cellular automata would be essential both for understanding complex systems in nature and for designing artificial complex systems" ([26] describing [40]). For these reasons, we select CAs as the primary domain for experimentation with our framework here.

In quantifying information storage, transfer and modification, we will of course be taking the Shannon perspective of **information theory** [41–43]. While this theory has traditionally been associated with signal processing, information is naturally the *language of computation*, and the theory has been increasingly used to study complex systems, e.g. [32, 34, 44, 45]. Importantly, the use of information theory means that our measures capture non-linear effects, and are *generic* and *portable* between application domains. This means that we can use the same measures and language to examine theoretical computing systems, computation in neural networks, gene networks computing their attractor, and many other systems.

We also highlight the importance of our focus on **the particular three operations** of information storage, transfer and modification as three *axes of complexity*. Our approach thus covers all of the distributed operations on information. Crucially, the approach is directly relevant to the concepts of memory, communication and processing which as we describe above are the terms with which computation is normally described. This means we can talk about distributed computation to subject matter experts in many disciplines, and concretely answer *meaningful questions* about their systems. For example, describing when and to what extent information is transferred between two brain regions, or where and how much information is stored in a given gene regulatory network. These types of concrete insights are more relevant to these subject matter experts than abstract notions of how complex their system is.

Finally, we emphasise the novelty and utility of our focus on the **local scale or dynamics** within the system. This means studying the manner in which information-theoretic averages are locally distributed in space and time in the dynamics of the system. Several authors have suggested that a complex system is better characterised by studies of its local dynamics than by averaged or overall measures (e.g. [39, 46]), yet the local dynamics of complex systems have received relatively little exploration. Certainly the dynamics of information storage, transfer and modification in distributed computation have not been explored. Quantifying and understanding distributed computation will necessitate studying these information dynamics and their interplay on a local scale in space and time. This will show how the computation unfolds

Table 1.1 Measures of information manipulation studied in the framework for the information dynamics of distributed computation

Measure	Concept under measurement
	Information storage (Chap. 3)
Excess entropy	Total storage
Active information storage*	Storage in use
(Temporal) entropy rate	Remaining uncertainty (complementary to storage)
	Information transfer (Chap. 4)
Apparent transfer entropy	Single source-destination transfer
Conditional transfer entropy*	Transfer considering other sources
Complete transfer entropy*	Transfer considering all other causal sources
Collective transfer entropy*	Transfer from a multivariate source
Multivariate transfer entropy*	Transfer from a multivariate source to a multivariate destination
Information flow	Causal effect
	Information modification (Chap. 5)
Separable information*	Non-trivial information modification
Information destruction*	Destroyed information

The original measures introduced in this thesis are marked with the symbol *. The primary measures used for each concept are shown in **bold**

through time, and the dynamics of how separate agents interact to achieve a collective task. Averaged or system-wide measures by definition cannot do this, and for these reasons we assert that the local perspective will provide greater insights into complex systems.

1.3 Contributions of this Thesis

In this thesis, we present the **first complete framework to quantify each of the information dynamics or component operations of distributed computation**. For each operation, we discuss a number of measures to quantify subtly different concepts (see Table 1.1), e.g. for information storage we separately consider total storage and the amount of storage that is actively in use. **Many of these measures are original contributions** (marked with * in Table 1.1); others have been previously introduced by other authors. For *all* of the measures, this thesis is the first presentation and application of them on a **local scale in space and time** in distributed computation, in order to specifically investigate the *dynamics of computation*. This is also the first quantitative investigation of **how the component operations interrelate** to produce distributed computation.

Importantly, our application of the local measures to cellular automata provides the **first quantitative evidence for long-held conjectures regarding the facilitation of distributed computation by emergent coherent structures in CAs**. That is, we show that blinkers are the dominant information storage entities, particles (gliders and domain walls) are the dominant information transfer agents, and particle collisions are the dominant information modification events. This demonstrates correspondence between our quantitative framework and the popularly-understood notions of memory, communication and processing. These results then have **significant implications for our fundamental understanding of distributed computation and the dynamics of complex systems**. This is because many other systems in nature (e.g. the opening and closing of stomatal apertures [47]) and those evolved to solve specific tasks (e.g. CAs evolved for a density classification task [48]) process information using similar emergent coherent structures. The application also demonstrates the large extent to which local measures provide insights that their averages cannot, with a particular use being filtering spatiotemporal dynamics to highlight coherent structure.

We also **resolve the relationships between key complementary concepts** for each operation, i.e.: total information storage and storage actively in use; information transfer and causal effect; and information modification and destruction. The distinctions revealed between information transfer and causal effect are particularly pertinent because of the large degree of confusion surrounding them in the literature (e.g. being directly equated in [49, 50]). The relationships could only be revealed using the appropriate quantification of the concepts identified here, coupled with our focus on local dynamics.

Furthermore, we use the framework to provide quantitative evidence for the conjecture that the **computational capabilities** of information storage and (coherent) transfer **are maximised near the critical or complex state** in *certain* order-chaos phase transitions.[2] Notably, we reconcile conflicting views of whether information transfer is maximised or at an intermediate level in such transitions by examining two complementary measures. Specifically, we find that information transfer is observed to be maximised in our experiments if one examines sources in isolation, but is observed to grow into the chaotic regime if one accounts for the interaction of sources (see apparent and complete transfer entropy respectively in Table 1.1). Importantly however, we show that such maximisations are not universally the case (in alignment with [22]), in particular for more complicated "transitions".[3]

Observing that coherent information structure (e.g. particles in CAs) is a defining feature of complex computation, we present a **methodology for studying coherent information structure**. This consists of state-space diagrams of the local information dynamics and a measure of structure in these diagrams. Crucially, we note that identifying coherent structure requires the local perspective. The methodology identifies both clear and "hidden" coherent structure in complex computation, most notably reconciling conflicting interpretations of the complexity in CA rule 22.

We demonstrate the utility of the framework in a number of crucial application areas. We study **computation in networks** using several models: random Boolean

[2] The type of phase transitions we are interested in are those with respect to a single order-chaos order parameter. In particular, we examine the *approximate* phase transitions in finite-size random Boolean networks as a function of connectivity. These networks are known to exhibit a true discontinuous phase transition in the infinite-size limit [51].

[3] An example here is CAs, where despite much conjecture there is no known smooth single-parameter phase transition between ordered and chaotic rules.

networks or RBNs (models of gene regulatory networks, GRNs), and cascading failures. We find maximisations of the information dynamics through order-chaos phase transitions in these systems, in alignment with popular conjectures, and find relationships between information dynamics and underlying topological structure. This is a particularly important application, since current opinion suggests that the next breakthroughs in network science will require understanding dynamics on networks [52–56]. This application demonstrates that information dynamics is a key candidate framework in this domain.

We also demonstrate the applicability of information dynamics to **computational neuroscience**, which is significant since understanding computation in the brain is a holy grail of complex systems science. In particular, we present a method for inferring directional, interregional information structure in multivariate data sets. The method is novel in providing the combined features of directional, non-linear, model-free analysis, which detects collective interactions, is applicable at the regional level and to relatively small data sets. We demonstrate the utility of the method by applying it to an fMRI data set, finding a tiered information structure which correlates well with the cognitive task the subjects were performing. This successful application indicates much scope for future investigation of information dynamics in this domain.

Finally, we present initial findings from using information dynamics to **guide the design of self-organised systems**. We demonstrate that this approach can result in the emergence of useful coherent information structure. These results could only be interpreted using our highly-illustrative investigation of computation in cellular automata. The results provide impetus for further investigation of how to harness our framework for system design.

1.4 Structure of the Thesis

We begin by discussing the current state of knowledge regarding computation in complex systems in Chap. 2. This chapter is intended to provide relevant background material on the two main theoretical bases of our approach, complex systems science and information theory, as well as to provide a deeper understanding of our motivation to quantify the information dynamics of distributed computation. Of primary importance, we discuss computation in cellular automata, the most significant complex system for exploring theoretical notions of distributed computation. This discussion establishes current qualitative understanding of computation in complex systems. We also discuss computation in networks, and its relevance to guiding self-organisation.

Subsequently in Chaps. 3–5, we consider each component operation on information in turn, and describe how to quantify each locally in a spatiotemporal system. In order to demonstrate the alignment of these measures with popularly-understood qualitative notions of computation, we measure each of these information dynamics at every point in space-time in several important CAs. This also demonstrates that the local measures provide useful spatiotemporal filters for coherent structure.

The first operation explored is information storage in Chap. 3. Here, we focus on measures of total information storage and storage actively in use in a computation. We provide the first quantitative evidence that blinkers (and to a lesser extent background domains) are the dominant information storage structures in CAs.

In Chap. 4, we describe how to quantify information transfer, in particular demonstrating subtle differences in the notion depending on whether one considers how the source couples with other sources in acting on the information destination. We provide the first quantitative evidence that particles in CAs are the dominant information transfer entities. We also clarify the relationship between the similar concepts of information transfer and causal effect.

We then describe the manner in which information storage and transfer are combined in the operation of information modification in Chap. 5. Significantly, we provide the first quantitative evidence that particle collisions in CAs are the dominant non-trivial information modification events. We also demonstrate the distinction between the concepts of information destruction and modification.

With the measures of the framework in place, we begin to apply it to several salient examples in Chaps. 6–8.

In Chap. 6, we measure the information dynamics through phase transitions in several notable network models: random Boolean networks (RBNs), and a model of cascading failures. We report maximisations of information storage and (coherent) transfer near the critical phase, in alignment with much conjecture on this topic.

Subsequently, in Chap. 7 we observe that such maximisation is not universally the case, but conjecture that *coherent information structure* is a defining feature of complex computation. We present a methodology for using information dynamics to study coherent information structure, and show that this methodology identifies both clear and "hidden" coherent structure in complex computation.

We then apply the framework to a number of biological and bio-inspired data sets in Chap. 8. This is done to illustrate the utility of the framework, and in particular the local perspective of information dynamics. We use information dynamics to develop a method for inferring directional, interregional information structure in brain imaging (fMRI) data. Also, we demonstrate the utility of information transfer as a generic fitness function in the domain of guided self-organisation, by showing that it leads to the emergence of coherent information structure (akin to particles in CAs).

Finally, we summarise the insights gained from this thesis in Chap. 9 and discuss directions for extending our explorations of the information dynamics of complex systems in the future.

References

1. M. Gell-Mann, *The Quark and the Jaguar* (W.H. Freeman, New York, 1994)
2. M. Prokopenko, V. Gerasimov, I. Tanev, Evolving spatiotemporal coordination in a modular robotic system, in *Proceedings of the Ninth International Conference on the Simulation of Adaptive Behavior (SAB'06)*, ed. by S. Nolfi, G. Baldassarre, R. Calabretta, J. Hallam,

D. Marocco, J.-A. Meyer, D. Parisi, Rome, ser. Lecture Notes in Artificial Intelligence, vol. 4095 (Springer, 2006), pp. 548–559.
3. K.I. Goh, A.L. Barabási, Burstiness and memory in complex systems. Europhys. Lett. **81**(4), 48002 (2008)
4. R. Morgado, M. Cieśla, L. Longa, F.A. Oliveira, Synchronization in the presence of memory. Europhys. Lett. **79**(1), 10002 (2007)
5. J.A. Brown, J.A. Tuszynski, A review of the ferroelectric model of microtubules. Ferroelectr. **220**, 141–156 (1999)
6. J. Pahle, A.K. Green, C.J. Dixon, U. Kummer, Information transfer in signaling pathways: a study using coupled simulated and experimental data. BMC Bioinform. **9**, 139 (2008)
7. I. Couzin, R. James, D. Croft, J. Krause, Social organization and information transfer in schooling fishes, in Fish Cognition and Behavior, ser. Fish and Aquatic Resources, ed. by B.C.K. Laland, J. Krause (Blackwell Publishing, Cambridge, 2006), pp. 166–185.
8. A.S. Klyubin, D. Polani, C.L. Nehaniv, All else being equal be empowered, in *Proceedings of the 8th European Conference on Artificial Life (ECAL)*, ed. by M.S. Capcarrere, A.A. Freitas, P.J. Bentley, C.G. Johnson, J. Timmis, U.K. Kent, ser. Lecture Notes in Computer Science. vol. 3630, (Springer, Berlin, 2005), pp. 744–753
9. M. Lungarella, O. Sporns, Mapping information flow in sensorimotor networks. PLoS Comput. Biol. **2**(10), e144 (2006)
10. O. Kinouchi, M. Copelli, Optimal dynamical range of excitable networks at criticality. Nat. Phys. **2**(5), 348–351 (2006)
11. J.J. Atick, Could information theory provide an ecological theory of sensory processing? Netw. Comput. Neural Syst. **3**(2), 213 (1992)
12. M.A. Sánchez-Montañés, F.J. Corbacho, Towards a new information processing measure for neural computation, ed. by J. Dorronsoro. in *Proceedings of the International Conference on Artificial Neural Networks (ICANN 2002)*, Madrid, 2002. Lecture Notes in Computer Science, vol. 2415 (Springer, Berlin, 2002), pp. 637–642.
13. T. Yamada, K. Aihara, Spatio-temporal complex dynamics and computation in chaotic neural networks, in *Proceedings of the IEEE Symposium on Emerging Technologies and Factory Automation (ETFA'94)*, Tokyo, 1994, pp. 239–244.
14. M.H. Jakubowski, K. Steiglitz, R. Squier, Information transfer between solitary waves in the saturable Schrödinger equation. Phys. Rev. E **56**(6), 7267 (1997)
15. A. Adamatzky (ed.), *Collision-Based Computing* (Springer, London, 2002)
16. D.E. Edmundson, R.H. Enns, Fully 3-dimensional collisions of bistable light bullets. Opt. Lett. **18**, 1609–1611 (1993)
17. S. Wolfram, *A New Kind of Science* (Wolfram Media, Champaign, 2002)
18. P. Gong, C. van Leeuwen, Distributed dynamical computation in neural circuits with propagating coherent activity patterns. PLoS Comput. Biol. **5**(12), e1000611 (2009)
19. P. Fernández, R.V. Solé, The role of computation in complex regulatory networks, in *Scale-free Networks and Genome Biology*, ed. by E.V. Koonin, Y.I. Wolf, G.P. Karev (Landes Bioscience, Georgetown, 2006), pp. 206–225
20. O. Miramontes, Order-disorder transitions in the behavior of ant societies. Complexity **1**(3), 56–60 (1995)
21. J.P. Crutchfield, The calculi of emergence: computation, dynamics and induction. Phys. D **75**(1–3), 11–54 (1994)
22. D.P. Feldman, C.S. McTague, J.P. Crutchfield, The organization of intrinsic computation: complexity-entropy diagrams and the diversity of natural information processing. Chaos **18**(4), 043106 (2008)
23. S. Lloyd, *Programming the Universe* (Vintage Books, New York, 2006)
24. K. Wiesner, M. Gu, E. Rieper, V. Vedral, Information erasure lurking behind measures of complexity, 2009, arXiv:0905.2918v1. http://arxiv.org/abs/0905.2918
25. C.G. Langton, Computation at the edge of chaos: phase transitions and emergent computation. Phys. D **42**(1–3), 12–37 (1990)

26. M. Mitchell, Computation in cellular automata: a selected review, in *Non-Standard Computation*, ed. by T. Gramss, S. Bornholdt, M. Gross, M. Mitchell, T. Pellizzari (Verlagsgesellschaft, Weinheim, 1998), pp. 95–140
27. M. Mitchell, A complex-systems perspective on the "computation vs. dynamics" debate in cognitive science, in *Proceedings of the 20th Annual Conference of the Cognitive Science Society (Cogsci98)*, Madison, 1998, ed. by M.A. Gernsbacher, S.J. Derry, pp. 710–715.
28. S. Wolfram, Universality and complexity in cellular automata. Phys. D **10**(1–2), 1–35 (1984)
29. J.L. Casti, Chaos, Gödel and truth, in *Beyond Belief: Randomness, Prediction and Explanation in Science*, ed. by J.L. Casti, A. Karlqvist (CRC Press, Boca Raton, 1991), pp. 280–327
30. R.V. Solé, S. Valverde, Information transfer and phase transitions in a model of internet traffic. Physica A **289**(3–4), 595–605 (2001)
31. S.A. Kauffman, *The Origins of Order: Self-Organization and Selection in Evolution* (Oxford University Press, New York, 1993)
32. J.P. Crutchfield, K. Young, Inferring statistical complexity. Phys. Rev. Lett. **63**(2), 105 (1989)
33. C.R. Shalizi, *Causal architecture, complexity and self-organization in time series and cellular automata* (University of Wisconsin-Madison, Ph.D. Dissertation, 2001)
34. J.P. Crutchfield, D.P. Feldman, Regularities unseen, randomness observed: levels of entropy convergence. Chaos **13**(1), 25–54 (2003)
35. A. Wuensche, Classifying cellular automata automatically: finding gliders, filtering, and relating space-time patterns, attractor basins, and the Z parameter. Complex **4**(3), 47–66 (1999)
36. A. Lafusa, T. Bossomaier, Hyperplane localisation of self-replicating and other complex cellular automata rules, in *Proceedings of the 2005 IEEE Congress on Evolutionary Computation*, vol. 1 (IEEE Press, Edinburgh, 2005), pp. 844–849
37. M. Prokopenko, Guided self-organization. HFSP J. **3**(5), 287–289 (2009)
38. J.E. Hanson, J.P. Crutchfield, Computational mechanics of cellular automata: an example. Physica D **103**(1–4), 169–189 (1997)
39. C.R. Shalizi, R. Haslinger, J.-B. Rouquier, K.L. Klinkner, C. Moore, Automatic filters for the detection of coherent structure in spatiotemporal systems. Phys. Rev. E **73**(3), 036104 (2006)
40. J. Von Neumann, Theory of self-reproducing automata, ed. by A.W. Burks (University of Illinois Press, Urbana, 1966).
41. C.E. Shannon, A mathematical theory of communication. Bell Syst. Tech. J. **27**(379–423), 623–656 (1948)
42. T.M. Cover, J.A. Thomas, *Elements of Information Theory* (Wiley, New York, 1991)
43. D.J. MacKay, *Information Theory, Inference, and Learning Algorithms* (Cambridge University Press, Cambridge, 2003)
44. A.S. Klyubin, D. Polani, C.L. Nehaniv, Representations of space and time in the maximization of information flow in the perception-action loop. Neural Comput. **19**(9), 2387–2432 (2007)
45. M. Prokopenko, F. Boschietti, A.J. Ryan, An information-theoretic primer on complexity, self-organization, and emergence. Complex **15**(1), 11–28 (2009)
46. J.E. Hanson, J.P. Crutchfield, The attractor-basin portait of a cellular automaton. J Stat. Phys. **66**, 1415–1462 (1992)
47. D. Peak, J.D. West, S.M. Messinger, K.A. Mott, Evidence for complex, collective dynamics and emergent, distributed computation in plants. Proc. Natl. Acad. Sci. USA **101**(4), 918–922 (2004)
48. M. Mitchell, J.P. Crutchfield, P.T. Hraber, Evolving cellular automata to perform computations: mechanisms and impediments. Physica D **75**, 361–391 (1994)
49. K. Ishiguro, N. Otsu, M. Lungarella, Y. Kuniyoshi, Detecting direction of causal interactions between dynamically coupled signals. Phys. Rev. E **77**(2), 026216 (2008)
50. X.S. Liang, Information flow within stochastic dynamical systems. Phys. Rev. E **78**(3), 031113 (2008)
51. A.S. Ribeiro, S.A. Kauffman, J. Lloyd-Price, B. Samuelsson, J.E.S. Socolar, Mutual information in random Boolean models of regulatory networks. Phys. Rev. E **77**(1), 011901 (2008)
52. D.J. Watts, *Six Degrees: The Science of a Connected Age* (Norton, New York, 2003)

53. M. Mitchell, Complex systems: network thinking. Artif. Intell. **170**(18), 1194–1212 (2006)
54. M. Mitchell, *Complexity: A Guided Tour* (Oxford University Press, New York, 2009)
55. F. Schweitzer, G. Fagiolo, D. Sornette, F. Vega-Redondo, A. Vespignani, D.R. White, Economic networks: the new challenges. Science **325**(5939), 422–425 (2009)
56. A.-L. Barabási, Scale-free networks: a decade and beyond. Science **325**(5939), 412–413 (2009)

Chapter 2
Computation in Complex Systems

Complex systems science is the study of large collections of (generally simple) entities, where the global behaviour is a non-trivial result of the local interactions of the individual elements [1]. This approach seeks a fundamental understanding of how such collective behaviour results from these interactions between simple individuals. In particular, it seeks to gain and apply this understanding across many different disciplines, examining both natural and man-made systems as apparently diverse as insect colonies, the brain, the immune system, economies and the world wide web [2]. Complex behaviour is often described as incorporating elements of both order and disorder (or chaos), and these elements can be seen in all of the above, e.g. path-following (order) versus exploration (disorder) in ant foraging.

While no common framework has been established for analysis of time-series of dynamics in complex systems science [1], increasingly "the notion of computation is being imported to explain the behaviour of" these complex systems [2]. To the lay reader, the concept of computation may seem rather distinct from these fields, yet the application is well-reasoned since the interactions between individuals can be seen as communication or signalling, and the system as a whole as processing information in determining its collective behaviour via a distributed computation.

The most prominent example of the use of the notion of computation to explain complex behaviour is with respect to cellular automata (CAs). CAs are discrete dynamical systems which are known to support complex computation, and have been used to model complex systems in nature. Their computation is qualitatively understood in terms of emergent coherent structures which are widely accepted to embody information storage, transfer and modification. Indeed this understanding is used to explain the computational function of similar emergent structures in neural circuits [3] and in plants [4]. Crucially however, quantitative evidence for such understanding is notably absent.

A quantitative understanding of the dynamics of these operations on information could provide a common analytic framework for complex systems, and would have important ramifications for the whole field. For example, insights into network topology in nature (e.g. [5, 6]) have been some of the greatest contributions of complex

J. T. Lizier, *The Local Information Dynamics of Distributed Computation in Complex Systems*, Springer Theses, DOI: 10.1007/978-3-642-32952-4_2,
© Springer-Verlag Berlin Heidelberg 2013

systems science, however it is widely acknowledged [2, 7–9] that understanding the *dynamics* of networks is the "next frontier" [8]. In particular, "the main challenge is understanding the dynamics of the propagation of information...in networks, and how these networks process such information" [9]. Another prominent example are attempts to guide the design of self-organised systems using an understanding of how they structure information [10]. Here we suggest that the task under design can be considered as a distributed computation, with an opportunity then for a framework for information dynamics to provide quantitative insights to such design.

Information theory provides the logical platform for our investigations, since information is the language of computation. Indeed, information theory has proven successful in analysing complex systems [1], initially through studies of characterising order-disorder continua as well as measuring complexity. It has many important features (e.g. being mathematically abstract) that give it the potential to become a leading framework for analysis and design of complex systems. Yet what we consider to be the most important of these features, the ability to analyse the information dynamics of distributed computation, is not yet properly established.

This chapter is used to introduce the reader to the two foundations of our work, complex systems science in Sect. 2.1 and information theory in Sect. 2.2. We will introduce the basic information-theoretic concepts used here, in particular the approach to studying information dynamics on a local scale in space and time. The chapter is also used to strongly highlight the need for quantitative insights into the information dynamics of computation in complex systems, and to introduce several relevant models that are analysed in later chapters. In Sect. 2.3 we describe the current state of understanding of distributed computation in cellular automata, the most important domain for theoretical discussions of this concept. We introduce the reader to network science in Sect. 2.4, highlighting the opportunities for a theory of information dynamics to complement this understanding of static structure. As an illustration, we describe an important model used in later analysis: random Boolean networks (RBNs). Finally, we discuss the use of complex systems science and information theory to design self-organised systems in Sect. 2.5. These examples make the opportunity to contribute a *framework for the information dynamics of distributed computation in complex systems* abundantly clear.

2.1 Complex Systems

Complex systems science is concerned with the study of systems which have two key characteristics [1]:

- They are composed of many simple elements;
- The elements interact in a non-trivial fashion.

Non-triviality of the interaction is particularly important to this definition. Prokopenko et al. [1] state that systems with a huge number of components interacting *trivially* are the domain of statistical mechanics, while those with *precisely defined* and constrained interactions are the domains of fields such as chemistry and engineering. Complex systems science however is concerned with the overlap of these, where non-trivial interaction of a large number of elements leads to intricate non-linear dynamics, violating classical (i.e. linear) assumptions. Complex systems science is important because of the prevalence of these characteristics in natural and man-made systems, and the challenge they present to traditional linear analysis.

Typically, computer simulation is used as the primary tool for the study of such systems. Here *agent-based modelling* [11, 12] considers the elements of the system as black boxes whose behavioural rules we know but whose internal structure we do not care about. The global behaviour of the system is allowed to emerge from the interaction of these simulated agents, modelling such emergence in the real-world system. Agent-based modelling has been described as "the most important conceptual tool introduced by complexity science" [11].

In one sense, the title *complex* captures the *difficult* nature of analysis of these systems. In another sense, it also captures the essence of their *interesting* nature. This is because the interaction of the elements on a microscopic level can give rise to sophisticated organisation of the system at a macroscopic level. Put another way, they can exhibit global behaviour which is interesting but not an obvious consequence of the local interactions. This is referred to as *emergent behaviour* [2].

Some authors go further in their definitions for complex systems. For example, Mitchell [2] proposes a complex system to be: "a system in which large numbers of components with no central control and simple rules of operation give rise to complex collective behaviour, sophisticated information processing, and adaptation via learning or evolution." There is no question that these terms are an essential part of our understanding of complex systems, but the extent to which some of these terms (e.g. adaptation) are necessarily part of complex systems or complex behaviour is debatable [1]. In the next sections, we discuss some of these related concepts. Subsequently, we consider the motivation for studying complex systems in more depth.

2.1.1 *Order, Disorder and Phase Transitions*

The underlying nature of complex systems in non-linear dynamics provides a very strong link to chaos theory (e.g. see [13]); indeed this field can be seen as one of the forerunners to complex systems science.[1] Complex systems theory is subtly different from chaos theory however: while chaos theory focuses on apparent randomness arising from very simple systems, complex systems science focuses on emergent global behaviour or organisation from non-trivial distributed interactions.

Importantly, complex systems are typically described as combining elements of order and randomness[2] to create truly intricate behaviour [11]. For example,

---

[1] For introductions to the beginnings of complex systems science, see [2, 11, 12, 14].

[2] In the sense of chaotic behaviour.

economies involve regulation and a perception of rational behaviour at the same time as wide variation in individual behaviours and unforeseen fluctuations in market dynamics.

Indeed, truly complex behaviour is neither completely ordered, nor completely random. Ordered systems are perfectly structured and therefore simple to predict [11]. Completely disordered systems cannot be predicted at all on an individual level, but prediction of average behaviour is not only possible but trivial. Complex systems on the other hand embody a *duality* between dependence and independence of their components, making prediction of them possible but non-trivial. Such observations are generalised by some authors in suggesting that complex behaviour occurs at a *phase transition* between ordered and chaotic behaviour (generalised in the *edge of chaos* hypothesis [15, 16]). Certainly there are many order-chaos phase transitions (in particular when they are dictated by a single order-chaos parameter) where complexity has been measured in some way to be maximised in the transition, e.g. maximisation of co-operative behaviour among ants at intermediate levels of ant density [17]. However, suggestions that such transition properties are universal, or that complex computation only occurs at such transitions in all systems[3] are strongly criticised in [18, 19].

2.1.2 Self-Organisation

The intricate link between the concepts of *self-organisation* and complex systems is shown by the phrasing "no central control" in Mitchell's definition [2] above.

A system is considered to be self-organised where it demonstrates two key features (see [20–22]):

- An increase in organisation (structure and/or functionality);
- Dynamics not guided by any centralised or external control agent.

These principles are generally accepted, although there remains some contention about their specific details. This includes how to quantify organisation (which differs between [20–22]), as well as the nature of external control and the extent to which it could be permitted.

Regardless of these specifics, it is clear that the global organisation in such systems result from distributed and localised interaction between the elements of the system; for example, in a school of fish individuals moderate their movement with reference to immediate neighbours rather than that of a central fish or of the whole school [23]. As such, self-organisation is a popular focus within complex systems science.

Related here is the concept *self-organised criticality*, where a system self-organises to a critical state near an order-chaos phase transition, and is stable to perturbations away from this state [24, 25]. *Scale-free* phenomena (or *fractal* or

---

[3] Particularly those with less well-defined or more complex transitions.

self-similar structure), where properties of agents or event sizes or inter-event distribution times follow a power-law, are a signature of these critical states.

2.1.3 Motivation for Studying Complex Systems

Motivations for studying complex and self-organised systems can be somewhat divided between *science*, or attempts to understand such systems, and *engineering*, or attempts to design or manipulate such systems for our own benefit.

Developing an understanding of complex systems is of immense importance, because such systems are widely distributed throughout nature. For example, Léveillé et al. state that understanding how "neurons cooperate to control behavioural processes is a fundamental problem in computational neuroscience" [26]. Other prominent examples include swarm behaviour [27], ant foraging [28], and heart beats [29]. Furthermore, complex behaviour arises in man-made systems, e.g. the internet [30] and city size distributions [25]. Complex systems science is very much an interdisciplinary field, both in terms of application domains and analytical approaches.

There are many open questions which require addressing before we can properly claim to have full understanding of the dynamics of these systems, including how such systems arise and how to classify types of systems and behaviour. Of prime importance is the common acceptance of how to quantify the concept of complexity, and sub-concepts such as self-organisation and emergence. Notions of computation are increasingly being used to describe complex systems; we will consider this perspective in relation to cellular automata in Sect. 2.3 (once we have established the required information-theoretical background). Also, fully understanding the dynamics of evolution (e.g. see [31]) is particularly important, because complex self-organised systems in nature have typically evolved over millions of years to address problems that are simple to state but hard to solve [28].

As such, engineering can and should take inspiration from these systems (e.g. as in [32]). This is particularly true because traditional engineering designs are finding their limitations in many situations: they are centralist, subject to single points of failure, have low tolerance to errors, are incapable of adapting to new situations, and scale poorly with problem size. As systems become more and more complex (with trends towards higher system densities and integration) new approaches need to be found. Sensor networks and robotic systems are two particular domains requiring innovative distributed approaches (e.g. [23, 33, 34]).

Not surprisingly then, in architecting solutions to address these issues, designers are looking to self-organised multi-agent systems. In general, this is because they display the key benefits of [23]:

- adaptability to change;
- robustness to failure and damage; and
- a high level of scalability.

Therefore, the ability to design self-organised systems is highly desirable in order to exploit these benefits. However as we describe in Sect. 2.5 most current approaches are ad-hoc (e.g. with genetic algorithms), either being specifically tailored to the given problem or not associated with a more widely used theoretical framework.

Indeed, this is part of the wider issue that complex systems science itself "lacks a common formal framework for analysis" [1]. The key goal of this science is to find general principles underlying classes of complex systems, and tools applicable across many systems from different domains. Certainly there have been successes here; arguably the greatest amongst these being the study of the topology or graph-theoretic properties of networks (see Sect. 2.4). With respect to time-series dynamics though, there is certainly a need for a widely accepted, rigorous and complete theoretical framework to be used. In the next section we describe information theory and its application to quantifying computation as a key candidate approach here.

2.2 Information Theory

Approaches considering computation and information are growing in popularity as candidate frameworks for the analysis and design of complex systems [1, 2, 35]. because commonalities in the handling of information Information theory is the language of computation and is at the centre of these approaches, which we describe in this section. We begin with an outline of the basic measures of information theory in Sect. 2.2.1. In Sect. 2.2.2 we then describe how these average measures can be extended to local dynamics in space and time, which is a key approach for this thesis. We also describe how to apply the measures to continuous variables in Sect. 2.2.3. We present support for the application of information theory to complex systems in Sect. 2.2.4, and describe several instances where information theory has already proven to be a useful framework for the design and analysis of complex systems. This leads us to highlight the opportunity for the establishment of a framework for quantifying the information dynamics of computation in complex systems.

2.2.1 Information-Theoretic Measures

Information theory was introduced by Shannon [36] to describe the transmission of information, incorporating concepts of reliability and efficiency. The basic concept of information theory is the communications channel, formed from an information source (which produces messages and encodes them onto a noisy and error prone channel) and a receiver (which decodes the channel's output in attempting to retrieve the original message). Information-theoretic quantities are formed by treating the specific messages x produced by the source as random variables X over a probability distribution function (PDF) $p(x)$ of the set of possible messages. Here we present the important information-theoretic quantities (see [37, 38] for further details). While we introduce the quantities in discrete form in this section, we describe extensions to continuous variables in Sect. 2.2.3.

The fundamental quantity is the Shannon **entropy**, which represents the average uncertainty associated with any measurement x of a random variable X (units in bits):

$$H_X = -\sum_x p(x) \log_2 p(x). \tag{2.1}$$

Specifically, this quantifies the *average* number of bits needed to encode the random variable X, and can thus be seen as the information content of the particular message. This information content, or the "informative" nature of a message, can be interpreted as how surprising or unlikely its event was [39]. The Shannon entropy can also be interpreted as the level of diversity in the source [1].

The **joint entropy** of two (or more) random variables X and Y is a generalisation to quantify the uncertainty of the joint distribution of X and Y:

$$H_{X,Y} = -\sum_{x,y} p(x, y) \log_2 p(x, y). \tag{2.2}$$

The **conditional entropy** of X given Y is the average uncertainty that remains about x when y is known:

$$H_{X|Y} = -\sum_{x,y} p(x, y) \log_2 p(x \mid y), \tag{2.3}$$

$$= H_{X,Y} - H_Y. \tag{2.4}$$

Note that we have:

$$p(x \mid y) = p(x, y)/p(y). \tag{2.5}$$

The **mutual information** between X and Y measures the average reduction in uncertainty about x that results from learning the value of y, or vice versa:

$$I_{X;Y} = \sum_{x,y} p(x, y) \log_2 \frac{p(x, y)}{p(x)p(y)}. \tag{2.6}$$

$$= H_X + H_Y - H_{X,Y}, \tag{2.7}$$

$$= H_X - H_{X|Y} = H_Y - H_{Y|X}. \tag{2.8}$$

From other perspectives, the mutual information can be said to measure how much information X and Y have in common, or how much information knowing the value of x tells one about the value of y on average. It can also be stated as the deviation of X and Y from independence (i.e. the Kullback–Leibler divergence of $p(x, y)$ from $p(x)p(y)$ [38]).

The mutual information can be generalised to a set of more than two variables as the **multi-information** or **integration** [40]. The multi-information is a measure of the deviation from independence of the G components in the system $\mathbf{X} = \{X_1, X_2, \ldots, X_G\}$:

$$I_{\mathbf{X}} = I_{X_1;X_2;\ldots;X_G} = \left(\sum_{g=1}^{G} H_{X_g} \right) - H_{X_1,X_2,\ldots,X_G}. \tag{2.9}$$

The multi-information of a set $\mathbf{Z} = \{\mathbf{X}, \mathbf{Y}\}$ can be expressed iteratively in terms of the multi-information of its components individually and the mutual information between those components:

$$I_{\mathbf{Z}} = I_{\mathbf{X}} + I_{\mathbf{Y}} + I_{\mathbf{X};\mathbf{Y}}. \tag{2.10}$$

The **conditional mutual information** between X and Y given Z is the mutual information between X and Y when Z is known:

$$I_{X;Y|Z} = \sum_{x,y,z} p(x, y, z) \log_2 \frac{p(x \mid y, z)}{p(x \mid z)}. \tag{2.11}$$

$$I_{X;Y|Z} = H_{X|Z} - H_{X|Y,Z} = H_{Y|Z} - H_{Y|X,Z}. \tag{2.12}$$

It can also be stated as the average common information between X and Y that was not contained in Z. This is the only valid "three-term entropy" [38] (though any of X, Y or Z can be joint variables). It is important to note that the three term expression can defy our intuition. In particular, though we might naively expect the conditional mutual information $I_{X;Y|Z}$ to always be smaller than the mutual information $I_{X;Y}$, it is also possible for $I_{X;Y|Z}$ to be larger. As described in [38], an example is where X, Y and Z are the input, noise and output from a binary symmetric channel. If the noise and input are independent $I_{X;Y} = 0$, but we will find that $I_{X;Y|Z} > 0$ because knowing the output provides us with some information about the previously unknown relationship between the input and noise.

The **channel capacity** is the maximum amount of information that a received signal Y can contain about a signal X transmitted through the channel. Since the information that the received signal contains about the transmitted signal is represented by their mutual information, channel capacity is specifically defined as the maximum mutual information for the channel over all distributions of the transmitted signal:

$$C(p(y \mid x)) = \max_{p(x)} I_{X;Y}. \tag{2.13}$$

This renders channel capacity as asymmetric and causal (in contrast to mutual information) [41]. Also in contrast to mutual information, it is a property of the channel itself rather than a property of the dynamics for a specific interaction over the channel.

The **entropy rate** is the limiting value of the rate of change of the joint entropy over k consecutive states of X, (i.e. measurements $x^{(k)}$ of the random variable $X^{(k)}$), as k increases [39]:

$$H_{\mu X} = \lim_{k\to\infty} \frac{H_{X^{(k)}}}{k} = \lim_{k\to\infty} H'_{\mu X}(k), \tag{2.14}$$

$$H'_{\mu X}(k) = \frac{H_{X^{(k)}}}{k}. \tag{2.15}$$

The entropy rate can also be expressed as the limiting value of the conditional entropy of the next state of X (i.e. measurements x_{n+1} of the random variable X') given knowledge of the previous k states of X (i.e. measurements $x_n^{(k)} = \{x_{n-k+1}, \ldots, x_{n-1}, x_n\}$, up to and including time step n, of the random variable $X^{(k)}$):

$$H_{\mu X} = \lim_{k\to\infty} H_{X'|X^{(k)}} = \lim_{k\to\infty} H_{\mu X}(k), \tag{2.16}$$

$$H_{\mu X}(k) = H_{X^{(k+1)}} - H_{X^{(k)}}. \tag{2.17}$$

We note that while these limiting values exist as $k \to \infty$ for stationary processes (e.g. see [37]), there is no guarantee that such limits exist for non-stationary processes.

Grassberger [42] first noticed that a slow approach of the entropy rate to its limiting value was a sign of complexity. Formally, Crutchfield and Feldman [39] use the conditional entropy form of the entropy rate (2.16)[4] to observe that at a finite block size k, the difference $H_{\mu X}(k) - H_{\mu X}$ represents the information carrying capacity in size k-blocks that is due to correlations. The sum over all k gives the total amount of structure in the system, quantified as **excess entropy**[5] (measured in bits):

$$E_X = \sum_{k=0}^{\infty} \left[H_{\mu X}(k) - H_{\mu X} \right]. \tag{2.18}$$

The excess entropy can also be formulated as the mutual information between the semi-infinite past and semi-infinite future of the system:

$$E_X = \lim_{k\to\infty} E_X(k), \tag{2.19}$$

$$E_X(k) = I_{X^{(k)};X^{(k^+)}} \tag{2.20}$$

where $X^{(k^+)}$ is the random variable (with measurements $x_{n+1}^{(k^+)} = \{x_{n+1}, x_{n+2}, \ldots, x_{n+k}\}$) referring to the k future states of X (from time step $n+1$ onwards). This interpretation is known as the **predictive information** [43], as it highlights that the excess entropy captures the information in a system's past which can be used to

---

4 $H_{\mu X}(k)$ here is equivalent to $h_\mu(L-1)$ in [39]. This means the sum in Eq. (2.18) starts from $k = 0$ as equivalent to $L = 1$.

5 The excess entropy was labelled the "effective measure complexity" by Grassberger in [42].

predict its future. This is significant as it is explicitly consistent with the interpretation of the excess entropy as the amount of structure or memory in the system.

2.2.2 Localising Information-Theoretical Measures

Information-theoretic variables are generally defined and used as an *average* uncertainty or information. We are interested in considering *local* information-theoretic values, i.e. the uncertainty or information associated with a *particular observation* of the variables rather than the average over all observations.[6] Local measures within a global average are known to provide important insights into the *dynamics* of non-linear systems [45]. Indeed, the ability to investigate *time-series dynamics* of complex systems provides an important connection from information theory to *dynamical systems theory* or *non-linear time-series analysis* (e.g. see [46, 47]). Importantly, only local information-theoretic measures can describe the *dynamics of computation*, since they alone can describe how information is being manipulated at each step in time. In this section we define how to obtain these local values and describe their meaning.

We use the mutual information as an illustrative example, though note that the derivation applies equally to the other terms defined in Sect. 2.2.1. The (average) mutual information defines the average information in X about Y or vice-versa; localisations consider how much information is conveyed by specific observations or realisations x_n and y_n of the variables X and Y at time step n.[7]

First, we note that the mutual information $I_{X;Y}$ is defined in Eq. (2.6) as a sum over all possible state tuple observations $\{x_n, y_n\}$, weighted by the probability $p(x_n, y_n)$ of observing each such tuple. This probability $p(x_n, y_n)$ is operationally equivalent to the ratio of the count of observations $c(x_n, y_n)$ of $\{x_n, y_n\}$, to the total number of observations N made: $p(x_n, y_n) = c(x_n, y_n)/N$. To precisely compute this probability, the ratio should be composed over all realisations of processes of the observed variables (as described in [50]); realistically however, estimates will be made from a finite number of observations. Subsequently, we replace the count by its definition

---

[6] *Local* information-theoretic measures are known as *point-wise* measures elsewhere [44].

[7] Appendix A describes two different approaches that have been presented to quantifying *partial* localisations of the mutual information $I(x_n; Y)$ [48]. The partial localisation $I(x_n; Y)$ considers how much information $I(x_n; Y)$ a specific value x_n at time step n gives about what value Y *might* take. We note this is distinct from the *full* localisations $i(x_n; y_n)$ that we consider here; this quantifies the amount of information conveyed by a specific value x_n about the specific value y_n that Y *actually takes* at time step n (or vice-versa). Our interest lies in these full localisations $i(x_n; y_n)$, as they quantify the specific amount of information involved or manipulated in the dynamics of the computation at time step n with the given realisation $\{x_n, y_n\}$. Appendix A demonstrates that there is only one approach to quantifying full localisations $i(x_n; y_n)$ that fulfils both additivity and symmetry properties.

We also note that a similar approach to "localising" information-theoretic values is by using sliding windows of observations (e.g. [49]). While this does provide a more local measure than averaging over all available observations, it is not local in the same sense as the term is used here (i.e. it does not look at the information involved in the computation at a *single specific time step*).

$c(x_n, y_n) = \sum_{g=1}^{c(x_n,y_n)} 1$, leaving the substitution $p(x_n, y_n) = \left(\sum_{g=1}^{c(x_n,y_n)} 1\right)/N$ into Eq. (2.6):

$$I_{X;Y} = \frac{1}{N} \sum_{x_n, y_n} \left(\sum_{g=1}^{c(x_n,y_n)} 1 \right) \log_2 \frac{p(x_n, y_n)}{p(x_n)p(y_n)}. \tag{2.21}$$

The log term may then be brought inside this inner sum:

$$I_{X;Y} = \frac{1}{N} \sum_{x_n, y_n} \sum_{g=1}^{c(x_n,y_n)} \log_2 \frac{p(x_n, y_n)}{p(x_n)p(y_n)}. \tag{2.22}$$

This leaves a double sum running over each actual observation g for each possible tuple observation $\{x_n, y_n\}$. This is equivalent to a single sum over all N observations:

$$I_{X;Y} = \frac{1}{N} \sum_{n=1}^{N} \log_2 \frac{p(x_n, y_n)}{p(x_n)p(y_n)}. \tag{2.23}$$

It is clear then that the mutual information measure is an *average* (or expectation value) of a *local mutual information* at each observation:

$$I_{X;Y} = \langle i(x_n; y_n) \rangle_n ; \tag{2.24}$$

$$i(x_n; y_n) = \log_2 \frac{p(x_n, y_n)}{p(x_n)p(y_n)}. \tag{2.25}$$

Note that by convention we use lower-case symbols to denote local values throughout this thesis.

The measure is *local* in that it is defined at each time step n. This method of forming a local information-theoretic measure by extracting the log term from a globally averaged measure is applicable to any of the aforementioned information-theoretic variables. For example, the conditional mutual information $I_{X;Y|Z}$ in Eq. (2.11) can also be expressed as the average of a *local conditional mutual information* $i(x_n; y_n \mid z_n)$ at each observation n:

$$I_{X;Y|Z} = \langle i(x_n; y_n \mid z_n) \rangle_n ; \tag{2.26}$$

$$i(x_n; y_n \mid z_n) = \log_2 \frac{p(x_n \mid y_n, z_n)}{p(x_n \mid z_n)}. \tag{2.27}$$

We also note that local mutual information values (including conditional ones) can be negative. This occurs where in Eq. (2.25) for example, $p(x_n, y_n) < p(x_n)p(y_n)$; i.e. there is more uncertainty $p(x_n \mid y_n)$ in x_n given y_n than there was uncertainty $p(x_n)$ in x_n independently of knowing y_n. These negative values are actually quite meaningful, and can be interpreted as there being negative information in the value of y_n about x_n. We could also interpret the value y_n as being *misleading* or *misinformative* about the value of x_n, because it had *lowered* our expectation of observing x_n prior to that observation being made in this instance. Importantly, it is not possible for local entropies $h(x_n) = -\log_2 p(x_n)$ to become negative.

The technique has been used (less explicitly) for the local excess entropy [50], the local statistical complexity [50, 51], and the local information [52]. Despite some interest from these authors [50–52], relatively little exploration has been made into the dynamics of these local information measures in complex systems, and certainly none has been made into the local dynamics of information storage, transfer and modification.

2.2.3 Information-Theoretic Measures of Continuous Variables

The information-theoretic measures in Sect. 2.2.1 are defined for discrete variables X and Y. In the discrete domain, measuring the relevant PDFs $p(x)$, $p(y)$ and $p(x, y)$ from a set of observations $\{x, y\} \in \{X, Y\}$ is straightforward. This is not the case for continuous variables X and Y. While it is possible to define the information-theoretic measures in integral form (see Chap. 9 of [37]), this requires the PDFs to be well-defined on the range of X and Y which is not the case for a finite number of samples.

A simple approach to handling continuous variables is to discretise the observations. One can then apply the relevant standard information-theoretic calculation to the discretised values. However with a slight increase in effort, one can remain in the continuous regime and so maximise the incorporation of the subtle features of the data into the calculations.

To do so in computing the transfer entropy[8] Schreiber [53] recommends using *kernel estimation* to estimate the required probabilities. This recommendation was used for example to compute transfer entropy in signal transduction by calcium ions in [54]. Kernel estimation relies on the notion that the given measure is *necessarily* an average of the local values of the measure in time (as per Sect. 2.2.2) rather than over all possible state transition tuples [47, 53]. First, the approach assumes that the required PDFs can be defined for each observation $n \in [1, N]$ in our set. For example, for the mutual information $I_{X;Y}$ the approach assumes that the PDFs for each observation ($\hat{p}_r(x_n, y_n)$, $\hat{p}_r(x_n)$ and $\hat{p}_r(y_n)$) can be defined, with the measure computed as an average over local values:

$$I_{X;Y} = \frac{1}{N} \sum_{n=1}^{N} \log_2 \frac{\hat{p}_r(x_n, y_n)}{\hat{p}_r(x_n)\hat{p}_r(y_n)}. \tag{2.28}$$

[8] The transfer entropy is arguably the most important measure used in this thesis. As such, recommendations on approaches to compute it will be followed for other measures also for consistency. The transfer entropy will be introduced in Chap. 4

The method then computes the probabilities for each observation $n \in [1, N]$ by counting the number of "similar" observations; the similarities are calculated using a kernel function Θ to judge "similarity" and a resolution r. For example, the probability of an observation (x_n, y_n) is estimated as:

$$\hat{p}_r(x_n, y_n) = \frac{1}{N} \sum_{n'=1}^{N} \Theta \left(\left| \begin{pmatrix} x_n - x_{n'} \\ y_n - y_{n'} \end{pmatrix} \right| - r \right). \tag{2.29}$$

By default Θ is the step kernel ($\Theta(x > 0) = 0$, $\Theta(x \leq 0) = 1$), and the norm $| \cdot |$ is the maximum distance. This combination results in $\hat{p}_r(x_n, y_n)$ being the proportion of the N values which fall within r of $\{x_n, y_n\}$ in both dimensions X and Y. Other choices for the kernel Θ and the norm $| \cdot |$ are possible. Conditional probabilities may be defined in terms of their component probabilities following Eq. (2.5). Importantly, a mutual information computed by kernel estimation is a slightly different quantity to one measured on discrete values: here the quantity measures the average reduction in uncertainty about predicting x *within* r that results from learning the value of y *within* r.

We note the existence of various techniques that attempt to improve upon kernel estimation, e.g. introducing optimisations in time or attempting to reduce errors. The most relevant of these are the techniques of Kraskov et al. [55, 56] for estimating the mutual information. This approach uses two specific enhancements designed to reduce errors when handling a small number of observations. The first is the use of Kozachenko-Leonenko estimates [57] of log-probabilities via nearest-neighbour counting; the second is to use a fixed number K of nearest-neighbours in the joint probability space (see [55, 56] for further details). This can be thought of as a kernel-estimation type technique, with a dynamically altered kernel width to adjust to the density of samples in the vicinity of any given observation. These adjustments smooth out errors in the PDF estimation. They also alter the meaning of the measure to being the information one learns from y about predicting the value of x within the nearest K values of $\{x, y\}$.

2.2.4 Reasons for Application to Complex Systems

Information theory is gaining popularity as an analytic tool for complex systems [1, 2, 58]. At first glance, obvious areas of its applicability lie in characterising order and disorder, since these have clear parallels to the concept of entropy [2]. The reasons for its application go beyond this however.

This highly *abstract nature* of information theory means that it makes no assumptions about the system. The only requirement for the application of information theory is *probability distributions* [21], which are generally accessible as they are a natural way of describing the dynamics of complex systems. This also means information-theoretic tools can handle *stochastic* as well as deterministic systems

[21]. The *generality* offered by information theory is crucial in complex systems analysis, where systems from all realms and varieties of science are candidates for investigation. Such generality means that analytic tools developed for one system are applicable to other superficially unrelated systems; this is a key goal of complex systems science.

Furthermore, information theory provides definitions which can be formulated mathematically, and its theoretical basis is well-developed, sound and stable [1]. Also, it captures *non-linear* relationships which we know to be a vital feature for any tool in complex systems science. Prokopenko et al. [1] claim that application of information theory to complex systems simply involves identification of the appropriate information channels, followed by computation of the relevant measures.

We must however note some arguments against the application of information theory here. Gibson [59] claimed that information theory was not applicable to natural systems as the environment does not send messages to living systems or agents. It is generally accepted now however that information theory does not need *explicit* messages on which to operate; its applicability comes in examining probability distributions of agent states and *implicit* messaging or influences. This is akin to the notion of *intrinsic computation* [19, 60]. Another criticism is that information-theoretic measures typically require a large amount of data in order to be accurately calculated. This is true to an extent, however modern computing tools (e.g. PCs) can handle typical data sets and generate meaningful results with ease, and techniques such as those of Kraskov et al. [55, 56] discussed in Sect. 2.2.3 have reduced the data size requirements.

From our perspective, information is the language of computation, so any attempt to characterise the dynamics of distributed computation, or to describe complex systems in these terms, must necessarily involve information theory. This is best captured by Gell-Mann [35]: "Although (complex adaptive systems) differ widely in their physical attributes, they resemble one another in the way they handle information. That common feature is perhaps the best starting point for exploring how they operate." This statement alludes to three important features of the information-theoretical perspective: its generality, focus on computation, and the known useful insights it produces. We will discuss the known useful insights it produces in the next sub-sections.

2.2.4.1 Examples of Information-Theoretic Analysis of Complex Systems

Information theory has already proven to be a useful framework for the design and analysis of complex self-organised systems [1]. As suggested above, it has been used to *characterise order and disorder*. Obviously the Shannon entropy can be used for this purpose, but this can be misleading (e.g. returning high values for periodic processes). More sophisticated is the use of the entropy rate in order to measure uncertainty in the context of the system's past, e.g. [19, 39].

Information theory has also been applied to the notoriously difficult task of attempting to *measure complexity* [1]. Various perspectives used here include the amount of information required to describe or predict the system or time required to implement it, or richness of the dynamics. General intuition is that completely statically ordered processes (with fixed states) and completely disordered or random process (with completely unpredictable states) have least complexity [11] and most perspectives aim to capture this. One common theme is the uncertainty (entropy) of system specific measures (e.g. perturbation avalanches in boolean networks [61]), or more generally the variance in information-theoretic measures (e.g. the variance of the *input entropy* with time [62]). Also, the Tononi-Sporns-Edelman (TSE) complexity [40] measures the extent to which a set of variables exhibit both global integration and at the same time functional segregation. It can be analytically related to spatial formulations of the excess entropy [63, 64].

Perhaps most prominent though is the *statistical complexity* [20, 65, 66]. Found in the study of *computational mechanics*, it attempts to capture the computational effort required to model complex behaviour. Formally, it is the uncertainty in the internal state of a minimal state machine (known as an ϵ-machine) that can statistically mimic a given process. This internal state is known as a *causal state*; two (possibly observationally different) states of a system have an equivalent causal state if the PDF of their future states is statistically equivalent. The ϵ-machine can be constructed for single agents, for system-wide collections of agents, or with a hybrid approach known as the *light-cone formulation*. The light-cone formulation forms causal states between past and future light cones centred at a given time step for a given agent. A *past light-cone* is the set of states of that agent and others that are causal ancestors of the given space-time point. A *future light-cone* contains the corresponding causal descendants.

A popular application of information theory has been the *investigation of order-chaos phase transitions* and whether measures of complexity are maximised in these transitions. These are effectively quantitative studies of the *edge of chaos* hypothesis [15, 16], which suggests that complexity is maximised in order-chaos phase transitions and in particular that computational properties are maximised there. For example, information-theoretic measures of complexity are demonstrated to be maximised at a critical phase in the construction of self-organised impact boundaries in [32], reflecting phase transitions established using task-specific measures. Other examples that suggest maximisations of information-theoretic measures of complexity in such phase transitions include [17, 67]. Some of the most relevant work regarding of the edge of chaos hypothesis considers cellular automata (these are discussed in Sect. 2.3.3). Importantly, other information-theoretic studies [19, 68] have demonstrated that there is no *universal* complexity trend through such phase transitions (in particular for more complicated phase transitions), and that computation can and does take place elsewhere on the order-chaos continuum.

With measures for complexity in place, *measuring self-organisation* as a change in complexity follows its definition in Sect. 2.1.2 and is typical, e.g. [20–22]. This is done for example by Shalizi et al. [20] using the statistical complexity. Similarly, complexity measures are also useful in quantifying *emergence*, e.g. see [50, 66].

Information theory has also been applied to the analysis of topological structure. Although such structure is static and contains no time-series dynamics, the measures are made on "observations" of the structure at each node or link in the network. For example, the amount of information the degree of nodes on either end of a given link have in common is considered in [69, 70].

Also, information-theoretic measures are being increasingly used to guide the design of artificial self-organised systems. This approach is discussed in detail in Sect. 2.5.

2.2.4.2 Information Dynamics of Computation in Complex Systems

It is important to observe that the applications described above go well beyond measuring only order and disorder, but address the nature of computation in complex systems. Many focus on the overall complexity of such computation. Our interest is in the information dynamics supporting such complex computation: how information is *stored*, *transferred* and *modified* in the system. Such concepts are considered most prominently in cellular automata, the most important venue for discussions of computation in complex systems. As we will discuss in detail in Sect. 2.3, a clear *qualitative* understanding of the dynamics of information in cellular automata has been established, but never validated with *quantitative* measures.

The lack of a definitive framework for these operations on information has impacts beyond cellular automata and across complex systems science. Their importance is underlined in that each of these operations has been considered in various complex systems settings. Information storage is considered a crucial part of the dynamics of human brain networks [71], synchronisation between coupled systems [72], coordinated motion in modular robots [73], and in the dynamics of inter-event distribution times [74]. Information transfer is manifested in "information cascades" spreading across schools of fish [75], said to give rise to self-organisation via dipole-dipole interactions [76], and optimum efficiency of information transmission is said to underpin order-chaos phase transitions in ant foraging [17]. Information modification is a key operation in collision-based computing models (e.g. [77, 78], including soliton dynamics and collisions [79]), as well as in biological neural networks and models thereof [80–83]

However, the lack of clearly established measures for these information dynamics has led to unanswered speculation on their role in complex computation. This is particularly true regarding information transfer, e.g. the conflicting suggestions that transfer is maximised in complex dynamics at a phase transition between ordered and chaotic behaviour [17, 67], or alternatively is at an intermediate level with maximisation leading to chaos [15, 84]. This lack of clarity has also led to the concepts of information transfer and causal effect sometimes being unfortunately directly equated, e.g. [85–87]. It also extends to the study of dynamics on networks, where as we will discuss in Sect. 2.4 understanding information transfer "is one of the most important open problems in science" [2].

Certainly there are good candidate measures for some of these concepts, notably the excess entropy [39] for measuring information storage (see Eq. (2.18)) and transfer entropy for information transfer (to be discussed in Chap. 4). However, their local dynamics have not been investigated, nor have they been demonstrated to align with accepted instances of these concepts in the dynamics of computation (see computation in CAs in Sect. 2.3.3). While brief explorations of local information-theoretic values values have been made (see Sect. 2.2.2), there is no established approach to studying the information dynamics of computation that aligns with dynamical systems theory. In particular, we have no understanding of how the component operations of information dynamics relate to each other. What we seek here are not simply more complexity measures, but a methodology to explain the dynamics of how complex distributed computation occurs.

In the following sections, we discuss in more detail the need for a framework for information dynamics in a number of specific application domains: cellular automata, the dynamics of networks, and guided self-organisation.

2.3 Cellular Automata

Cellular automata (CAs) are an important general class of models, since they support complex computation and provide the ability to model complex systems in nature [88]. They are *the most important domain for the study of distributed computation*, as the subject of a large amount of related work regarding the nature of computation in complex systems (e.g. [15, 51, 52, 62, 88–95]). Significantly, Von Neumann was known to be a strong believer that "a general theory of computation in 'complex networks of automata' such as cellular automata would be essential both for understanding complex systems in nature and for designing artificial complex systems" ([88] describing [96]). We select CAs for experimentation here for these reasons and because there is very clear *qualitative* observation of emergent structures representing information storage, transfer and modification therein (e.g. [15, 88]).

In this section, we describe the mechanics of CAs in Sect. 2.3.1, then discuss interpretations of complex behaviour and emergent structures in CAs in Sect. 2.3.2. We explore the perspective of computation within CAs and outline opportunities for providing quantitative insights on the nature of such computation in Sect. 2.3.3. The reader is introduced to several of the important CA rules investigated in this thesis in Sect. 2.3.4. We also describe existing techniques for filtering emergent structure in CAs in Sect. 2.3.5, outlining how the ability to filter such structure from a computational perspective would provide a novel contribution.

2.3.1 Functionality of Cellular Automata

CAs are discrete dynamical lattice systems. They consist of an array of cells which each update their discrete state as a function of the states of a fixed number of spatially neighbouring cells using a uniform rule. These updates occur synchronously in discrete time. Although the behaviour of each individual cell is very simple, the (non-linear) interactions between all cells can lead to very intricate global behaviour. As such, CAs have become a classic example of self-organised complex behaviour. Of particular importance, CAs have been used to model real-world spatial dynamical processes, including fluid flow, earthquakes and biological pattern formation [88].

The neighbourhood used as inputs to a cell's update rule at each time step is usually some regular configuration. In 1D CAs, this means the same range r of cells on each side and includes the current state of the updating cell. One of the simplest variety of CAs—1D CAs using binary states, deterministic rules and $r = 1$ neighbour on either side—are known as the *Elementary CAs*, or *ECAs*. As such, the parent nodes which determine $x_{i,n+1}$ (the value of node X_i at time $n + 1$) in an ECA are $\{x_{i-1,n}, x_{i,n}, x_{i+1,n}\}$. The past light-cone of $x_{i,n+1}$ is then made up of these nodes, their parents, and so on. The future light-cone of $x_{i,n+1}$ consists of the nodes it has a direct causal effect on, i.e. $\{x_{i-1,n+2}, x_{i,n+2}, x_{i+1,n+2}\}$, the nodes they have a direct causal effect on, and so on.

Example evolutions of ECAs from random initial conditions may be seen in Fig. 2.1 for the examples we discuss in Sect. 2.3.4. For more complete definitions of CAs, including the definition of the Wolfram rule number convention for specifying update rules, see Ref. [92].

2.3.2 Complex Behaviour in Cellular Automata

Wolfram [89, 92] sought to classify CA rules in terms of their asymptotic behaviour in time. He proposed four classes of asymptotic behaviour: I. Homogeneous state; II. Simple stable or periodic structures; III. Chaotic aperiodic behaviour; IV. Complicated localised structures, some propagating. This classification has been highly influential on subsequent research due to the parallels (for discrete state and time systems) with our knowledge of dynamical systems. Here classes I and II represent *ordered* behaviour, and class III represents *chaotic* behaviour. Class IV represents *complex* behaviour and is considered to lie between the ordered and chaotic classes.

The analogy to dynamical systems is taken further in considering the *state-space* of the global state $\mathbf{X}$ of the CA, and the *attractors*[9] and *transient* paths[10] in this space [62]. This discrete state-space is analogous to Poincaré's *phase portrait* in continuous dynamics. Ordered CAs are said to exhibit short transient paths with high convergence to attractors. Chaotic CAs exhibit long transient paths with low

[9] An attractor is a single global state $\mathbf{x}_n = \{\ldots, x_{i-1,n}, x_{i,n}, x_{i+1,n}, \ldots\}$ or periodic sequence of global states that a CA (generally considering fixed sizes) can reach after a finite number of time steps but then never leave (unless some stochasticity is introduced into the dynamics).

[10] A transient is a path of global states that a CA could traverse before reaching an attractor. For deterministic CAs, two or more transient paths can converge on the same next state, but a given state cannot diverge via multiple transient paths to more than one next state. A CA of finite size C cells must reach an attractor after a finite number of time steps (since there are a finite number b^C of possible global states, where b is the base or number of possible discrete states for each cell).

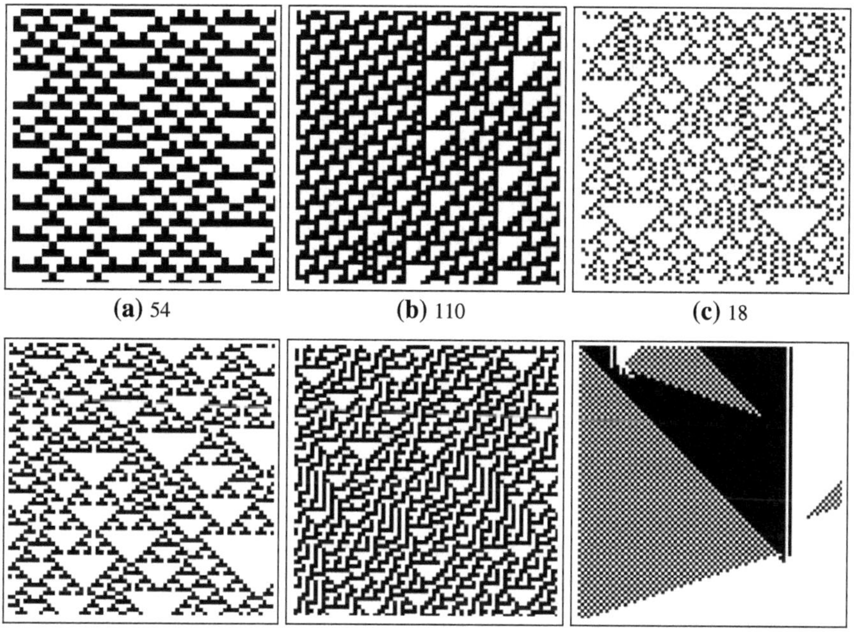

(a) 54 **(b)** 110 **(c)** 18 **(d)** 22 **(e)** 30 **(f)** ϕ

Fig. 2.1 Example time-evolutions of states of several important CA rules from random initial conditions. All rule numbers follow the Wolfram rule number convention [92], except rule ϕ which is an $r = 3$ CA evolved to classify the density of initial states (see Sect. 2.3.4). Black cells are in the "1" state, white cells are in the "0" state. Time increases down the page for all CA plots: one row represents one time slice $\mathbf{x}_n$ of the states of each cell in the CA. Note: the first row shown here is typically for some time $n > 0$ (NB: (a)–(c) Reprinted with permission from J. T. Lizier et al. [97]. Copyright 2010, American Institute of Physics. (d)–(e) are reprinted from Ref. [98] with permission of Springer.)

convergence to attractors. In contrast, complex CAs are observed to exhibit maximal uncertainty in the length of transient paths.

Much conjecture remains as to whether Wolfram's classes are quantitatively distinguishable however (e.g. see [99]). Regardless of this however they certainly provide an interesting qualitative analogy to dynamical systems.

More importantly, the approach seeks to characterise complex behaviour in terms of *emergent structure* in CAs: *particles*, *gliders*, *blinkers*, *domains* and *domain walls*. Qualitatively, a domain may described as a set of background configurations in a CA, for which any given spatially-extended configuration will update to another such configuration in the set in the absence of any disturbance. Domains are formally defined within computational mechanics [94] as spatial process languages in the CA. Particles are qualitatively considered to be elements of *coherent* spatiotemporal structure which disturb background domains. Gliders are particles which repeat periodically in time while moving spatially, while repetitive non-moving particle

(a) (b)

Fig. 2.2 Example of filtering of CA rule 110: **a** raw states; **b** local statistical complexity [51] of these states, which clearly highlights gliders and blinkers against the background domain. These diagrams were produced using the CimulA package [100], then rotated to have cells across and time down the page (matching Fig. 2.1)

structures are known as *blinkers*.[11] Formally, particles are defined within computational mechanics as a boundary between two domain regions [94]; as such, they can also be termed as domain walls, though this is typically used with reference to aperiodic particles.

These emergent structures are more clearly visible when the CA is *filtered* in some way, e.g. see local statistical complexity [51] applied to ECA rule 110 in Fig. 2.2. We discuss techniques for such filtering in Sect. 2.3.4. First, we need to understand how distributed computation is studied and measured in CAs.

2.3.3 Computation in Cellular Automata

CAs can be interpreted as undertaking distributed computation: it is established that "data represented by initial configurations is processed by time evolution" [89]. Indeed, any time evolution of the CA represents an intrinsic computation [19, 60] in determining its own future and ultimate attractor state and phase, in the same way that the universe "computes itself" [101].

As such, computation in CAs has been a popular topic for study (see [88]), with a particular focus in observing or constructing (Turing) universal computation in certain CAs. An ability for universal computation is defined to be where "suitable initial

[11] Of course, blinkers can be considered to be vertical or non-moving gliders, while both are particles.

configurations can specify arbitrary algorithm procedures" in the distributed computing entity, which is capable of "evaluating any (computable) function" [89]. Wolfram conjectured that all class IV complex CAs were capable of universal computation [89, 90]. He went on to state that prediction in systems exhibiting universal computation is limited to explicit simulation of the system, as opposed to the availability of any simple formula or "short-cut" [89, 90]. He drew parallels to undecidability in the halting problem for universal Turing machines which are echoed by Langton [15] and Casti [102]. Casti extended the analogy to undecidable statements in formal systems (i.e. Gödel's Theorem). The undecidability is also linked back to the maximal uncertainty in transient lengths for complex or critical behaviour in other systems (e.g. see [103]). The capability for universal computation has been proven for several CA rules, through the design of rules generating elements to (or by identifying elements which) specifically provide the component operations required for universal computation: **information storage**, **transmission** and **modification**. Examples here include most notably the 2D rule known as the "Game of Life" [104] and ECA rule 110 [105]; also see [106] and discussions in [88].

The focus on elements providing information storage, transmission and modification pervades discussion of all types of computation in CAs (e.g. see also "collision-based computing" in [78, 107]). Wolfram claimed that in class III CAs information propagates over an infinite distance at a finite speed, while in class IV CAs information propagates irregularly over an infinite range [90]. Langton [15] hypothesised that complex behaviour in CAs at the *edge of chaos* exhibited the three component operations required for universal computation. He suggested that the more chaotic a system becomes the more information transmission increases, and the more ordered a system becomes the more information it stores. Complex behaviour was said to occur at a phase transition between these extremes requiring an *intermediate level* of both information storage and transmission. He suggested for example, that intermediate levels of information transmission can support large correlation lengths, but if information propagates too well, *coherent information decays into noise*.

Importantly, Langton elaborates [15] that transmission of information means that the "dynamics must provide for the propagation of information in the form of signals over arbitrarily long distances". He goes on to suggest that **particles form the basis of information transmission**, since they appear to facilitate communication about the dynamics in one area of the CA to another area. To complete the qualitative identification of the elements of computation in CAs, he also suggested that **blinkers formed the basis of information storage**, and **collisions** between propagating (particles) and static structures (blinkers) "**can modify either stored or transmitted information in the support of an overall computation**" or decision process about the dynamics. He also made rudimentary attempts at quantifying the average information transfer (and to some extent information storage), via mutual information.[12] Recognising *the importance of the emergent structures to computation*, several examples exist of attempts to automatically identify CA rules which give rise to particles

[12] However as discussed in Chap. 4 this is a symmetric measure not capturing directional transfer.

and gliders, e.g. [62, 108], suggesting these to be the most interesting and complex CA rules.

Several authors however criticise the aforementioned approaches of attempting to classify CAs in terms of their generic behaviour or "bulk statistical properties", suggesting that the wide range of differing dynamics taking place across the CA makes this problematic [88, 94]. Related here is that, despite Langton's hypothesis and the introduction of useful measures of the complexity of CA rules (e.g. [19, 51, 62]), there is no established phase transition dictated by a single order-chaos parameter in CAs [19]. Also, Gray suggests that there there may indeed be classes of CAs capable of more complex computation than universal computation alone [99]. More importantly, Hanson and Crutchfield [94] criticise the focus on universal computational ability as drawing away from the ability to identify "generic computational properties", i.e. a lack of ability for universal computation does not mean a CA is not undertaking any computation at all.

Alternatively, these studies suggest that analysing the rich space-time dynamics *within* the CA is a more appropriate focus. As such, references [88, 94] and others have analysed the *local* dynamics of intrinsic or other specific computation, focusing on the computational roles of emergent structures. They align with Langton's observations of blinkers storing information, particles facilitating the transfer of information and collisions facilitating the information modification or processing. Notable examples here include: the method of applying filters from the domain of computational mechanics by Hanson and Crutchfield [94]; and analysis using such filters to analyse CA rules selected via evolutionary computation to perform classification tasks by Mitchell et al. [93, 109]. Also relevant are studies which deeply investigate the nature of particles and their interactions, e.g.: particle types and their interaction products identified for particular CAs in [109–112], rules established for their interaction products in [113], and studies of "collision-based computing" in [78, 107].

This perspective of the computational roles of emergent structures is important not only for our theoretical understanding of the nature of distributed computation, but is also important on a practical level. This is because it has been used to explain the computational function of similar propagating coherent emergent structures and their interactions in neural circuits [3] and in the opening and closing of stomatal apertures in plants [4].

Despite such interest, there is no quantitative evidence to support these conjectures about the role of emergent structures in computation in CAs. Simply quantifying the average information dynamics in CAs would be novel itself; we are not aware of any previous direct measurement of information storage, transfer or modification in CAs.[13] However measuring *averages* will not suffice: e.g. a coincidence of

[13] Langton's use of mutual information in [15] is a symmetric measure not capturing directional transfer. As we will describe in Sect. 3.1, Grassberger inferred the excess entropy to be infinite in some CAs under certain circumstances by studying trends of the entropy rate [42, 114]. The study did not make any direct measurements of the excess entropy apart from these inferences, and focussed on collective excess entropy. Crutchfield and Feldman measure the excess entropy

high information transfer in CAs with gliders would not mean that gliders are the information transfer agents. Providing such evidence requires the ability to quantify the information dynamics of computation on a *local scale in space and time* within the CA. It is only the local scale that will explicitly show the computational roles of the emergent structures. Yet there is no complete framework that locally quantifies the individual information dynamics of distributed computation within CAs or other systems. In this thesis we will provide such a framework in Chaps. 3–5, and use to it to describe how the component operations of computation interact to give rise to emergent complex behaviour. We expect this framework to highlight blinkers and domain regions as dominant information storage processes, particles (including gliders and domain walls) as information transfer, and particle collisions as information modification. This will make a significant contribution to our fundamental understanding of the nature of distributed computation.

2.3.4 Examples of Distributed Computation in CAs

In this thesis, we will examine the computation carried out by several important ECA rules:

- Class IV complex rules 110 and 54 [92], both of which exhibit a number of glider types and collisions. See raw states in Fig. 2.1a, b, and a filtered view of the gliders in rule 110 in Fig. 2.2. ECA rule 110 is the only proven computationally universal ECA rule [105].
- Rules 22 and 30 as representative class III chaotic rules [92] (see Fig. 2.1d, e).
- Rules 18 as a class III rule which contains domain walls against a chaotic background domain [91, 94] (see Fig. 2.1c).

While these ECAs are not computing any particular human understandable task, the intrinsic computation they are undertaking is of interest.

We also examine a CA carrying out a "human-understandable" computational task. ϕ_{par} is a 1D CA with range $r = 3$ (Wolfram rule number 0xfeedffdec1aaeec0-eef000a0e1a020a0). It resulted from an evolutionary computation experiment by Mitchell et al. [93, 109] which aimed to classify whether an initial CA configuration had a majority of 1's or 0's by reaching a fixed-point attractor of all 1's for the former or all 0's for the latter. This CA rule achieved a success rate above 70 % in its task. An example run of this CA can be seen in Fig. 2.1f. The CA appears to have evolved to carry out this computation using blinkers and domains for information storage, gliders for information transfer and glider collisions for information modification. The computation has been interpreted as follows [93, 109]. The CA exhibits an initial emergence of domain regions of all 1's or all 0's storing information about local high densities of either state. Where these domains meet, a checkerboard domain

---

(Footnote 13 continued)
in *spatially*-extended blocks in various CAs in an ensemble study in [19], however this is not an information storage since the measurement is not made *temporally*.

propagates slowly (1 cell per time step) in both directions, with the gliders at the leading edge transferring information regarding a *soft* uncertainty about the density in this part of the CA. Some "certainty" is provided where the checkerboard encounters a blinker boundary between 0 and 1 domains, which stores information about a *hard* uncertainty in that region of the CA. This results in an information modification event where the domain on the opposite side of the blinker to the incoming checkerboard is concluded to represent the higher density state, and is allowed to propagate over the checkerboard domain. Because of the greater certainty attached to this decision, this new information transfer occurs at a faster speed (3 cells per time step); it can overrun checkerboard regions, and in fact collisions of opposing types of this strong propagation give rise to the (hard uncertainty) blinker boundaries in the first place. The final configuration is the result of the interactions in this distributed computation.

In all of these CAs we expect a framework for local information dynamics to quantitatively confirm these qualitative observations of the computational roles of emergent structure. This will provide a deeper understanding of computation than single or generic measures of bulk statistical behaviour, from which conflict often arises in attempts to provide classification of complex behaviour. In particular, we seek clarification on the long-standing debate regarding the nature of computation in ECA rule 22.

Suggestions that rule 22 is complex include the difficulty in estimating the metric entropy (i.e. temporal entropy rate) for rule 22 in [42], due to "complex long-range effects, similar to a critical phenomenon" [114]. This effectively corresponds to an implication that rule 22 has contains an infinite amount of collective memory (see Sect. 3.1). Also, from an initial condition of only a single "on" cell, rule 22 forms a pattern known as the "Sierpinski Gasket" [92] which exhibits clear self-similar structure. Furthermore, rule 22 is a 1D mapping of the 2D Game of Life CA (known to have the capability for universal computation [104]) and in this sense is referred to as "life in one dimension" [115], and complex structure in the language generated by iterations of rule 22 has been identified [116]. Also, we report here that we have investigated the C_1 complexity measure [117] (an enhanced version of the variance of the input entropy [62]) for all ECAs, and found rule 22 to clearly exhibit the largest value of this measure (0.78 bits to rule 110's 0.085 bits). Similarly, the statistical complexity [20, 51, 65] measure is larger for rule 22 than rules 54 and 110 (at 4.22, 3.80 and 3.91 bits respectively).[14]

On the other hand, suggestions that rule 22 is not complex include its high sensitivity to initial conditions and perturbations leading to Wolfram and Grassberger classifying it as class III chaotic [92, 114]. Gutowitz and Domain [95] claim this renders it as chaotic despite the subtle long-range effects it displays, further identifying its fast statistical convergence, and exponentially long and thin transients in state space (see [62]).

---

[14] Measured using the CimulA package [100] over 600 time steps of 100000 cells, with light cone depths of 3 time steps.

Importantly, no coherent structures (particles, collisions, etc.) are found for rule 22 using a number of filters (e.g. local statistical complexity [51]). This reflects the *paradigm shift to an examination of local dynamics* rather than generic, overall or averaged analysis. In our approach, we seek to combine this local viewpoint of the dynamics with a quantitative breakdown of the individual elements of computation, and will investigate the computation of rule 22 in this light.

2.3.5 Filtering Structure in Cellular Automata

Measuring the local information dynamics of computation at each space-time point in a CA will create spatiotemporal profiles which can be viewed as a method of filtering the CA. As previously suggested, several methods already exist for filtering the important structural elements (i.e. particles) in CAs [51, 52, 62, 94, 118, 119]. These will provide an important basis for comparison to our spatiotemporal profiles.

The earliest methods were hand-crafted for specific CAs (relying on the user knowing the period of background domains) by Grassberger [118, 119]. Later methods can be automatically applied to any given CA, including finite state transducers to recognise the regular spatial language of the CA using ϵ-machines by Hanson and Crutchfield [94, 120]. Helvik et al. use local information (i.e. local spatial entropy rate) [52]. Wuensche displays executing rules with the most frequently occurring rules filtered out [62]. Shalizi et al. use local statistical complexity (via the light-cone formulation) and local sensitivity [51]. All of these successfully highlight particles against the background domain. Note the use of several *local* information-theoretic measures here (since only local measures, not averages, provide values at every spatiotemporal point).

Obviously, filtering is not a new concept, however the ability to *separately* filter each of the information dynamics of computation would be novel. Most importantly, it would be the first quantitative study of the dynamics of each of these computational operations within the CA. In potentially separating the emergent structures representing each of the information dynamics, this would provide the first quantitative evidence for their computational roles. It would also show how these operations interrelate to give rise to complex behaviour, in comparison to other filters which give only a single view of where that complexity occurs. This would allow a more refined investigation than single measures, and should reveal interesting differences in the parts of the emergent structures that are highlighted. Furthermore, these measures will be generally applicable to any multivariate time-series, unlike some of the filtering measures here (e.g. spatial ϵ-machines [94, 120] and spatial entropy rate [52]) which are only applicable to lattice systems.[15]

[15] Lattice systems are those with a regular spatial ordering for their agents or variables, typically being placed on a one or two-dimensional array.

2.4 The Dynamics of Networks

Complex systems science has been particularly successful in the study of network topology (e.g. see [30, 121]), where the key concepts of small-world [5, 7] and scale-free [6, 122, 123] structures have attracted an enormous amount of attention. In particular, this is because both such topologies are found to be widespread amongst naturally occurring and man-made systems. Small-world topologies balance fixed and random network structures to provide both short path length and high clustering (or from another perspective, provide high global and local efficiency of communication [124]). That small-world networks represent a balance between fixed and random structures resonates with observations that complex systems balance ordered and chaotic properties (see Sect. 2.1.1). Scale-free topologies display a distribution of the degree of nodes (i.e. number of connections to other nodes) that is inversely proportional to the degree (sometimes called a $1/f$ distribution). The arising of scale-free topologies can be explained via the principle of "preferential attachment" [6, 122]—that new nodes introduced to the system preferentially make connections to nodes in proportion to their existing number of connections (because connections to well-connected nodes will prove more rewarding in some sense). A scale-free distribution is highly structured, and is considered to be a signature of self-organised criticality (see [24, 25]).

Yet the time-series behaviour or dynamics on networks have received less attention and are "much less well understood" [9]. There is a deep need for fundamental insights into network dynamics, and how these are related to the underlying structure [125]. Indeed Barabási, who introduced the concept of preferential attachment, states that "we need to tackle the next frontier, which is to understand the dynamics of the processes that take place on networks" [8]. Similarly Watts, who introduced the small-world networks concept, states that "next to the mysteries of dynamics on a network—whether it be epidemics of disease, cascading failure in power systems, or the outbreak of revolutions—the problem of networks that we have encountered up to now are just pebbles on the seashore" [7].

To some degree, the time-series dynamics of state-space trajectories and damage spreading are established, e.g. [126–129]. However, Barabási states that a major problem here is the diversity of types of dynamical process, and wonders whether "these dynamical processes share some common characteristics? I suspect that such commonalities do exist; we just have not yet found the framework to unveil their universality" [8]. He goes on to suggest that if such a framework were to be found, then "combined with the universality of the network topology, we may soon have something that could form the foundation of a theory of complexity" [8].

We believe that the abstract nature of information theory places it as an ideal candidate for investigating and comparing dynamics across different network types, and revealing their commonalities. In particular, the information dynamics of computation align with the way many authors talk about dynamics on networks. This is underlined by Mitchell [9] who suggests that "the main challenge is understanding the dynamics of the propagation of information…in networks, and how these networks process such information." Later, Mitchell also states that "understanding the ways in which information spreads in networks is one of the most important open problems in science" [2].

These views are echoed in several studies which have investigated the propagation and the processing of information in networks, in particular reporting maximisations of these properties at (approximate) phase transitions between ordered and chaotic regimes. Solé and Valverde [67] investigated the effect of varying the message generation rate in a model of computer networks, finding phase transitions maximising the number of packets actually delivered and the mutual information in the status of random node pairs. They infer that information transfer is maximised at the critical state. Kinouchi and Copelli [80] investigated varying the "branching ratio" (effectively an activity level) in a network of excitable elements, finding phase transitions maximising the dynamic range of the element's output, and inferring a maximisation of information processing at criticality. The spread of information was investigated in [130] under a number assumptions of average diffusion behaviour in a stochastic network model, finding better spread of information in scale-free rather than completely random networks. These investigations align with much conjecture regarding computational properties being maximised at the *edge of chaos* between ordered and chaotic behaviour, e.g. [15, 16, 103].

We are particularly interested in investigating the information dynamics of **random Boolean networks** (RBNs) ([16], and see [131]), in part because of the power in their generality as discrete dynamical network models with a large sample space available. Also, they have a well-known phase transition from ordered to chaotic dynamics, in terms of length of transients in phase space with respect to average connectivity or activity level. We are also motivated by their popularity as models of Gene Regulatory Networks (GRNs). Perhaps most importantly, there have been several recent attempts to study the computational properties of RBNs (in particular information transfer) since they are useful generalised models of computation in networks. Here, Ribeiro et al. [132, 133] measure mutual information in the states of random node pairs as a function of connectivity in the network, and Rämö et al. [61] measure the uncertainty (entropy) in the size of perturbation avalanches as a function of an order parameter. Both find maximisation near the critical point, claiming that their results imply maximisation of information propagation in this regime.

We are also interested in investigating the information dynamics of **cascading failure events** [134, 135]. These are local failures that trigger avalanche mechanisms with large effects over the whole network (e.g. catastrophic blackouts in power grids [134, 136] and global failure in financial markets [125, 137]). Similar to studies of damage spreading or perturbation avalanches on networks (e.g. [129, 138]), investigations of this concept consider how information spreads on the network; indeed: "the phenomena of cascading failures emphasises the need to understand information spreading and how it is affected by network structure" [2]. Often these cascades or avalanches are directly identified with information transfer (e.g. waves of directional change in schooling fish are referred to as "information cascades" [75]). Certainly information transfer is an integral part of cascade dynamics, but the relationship may not be as trivial as a one-to-one mapping. More generally, our interest lies in examining the way the network *intrinsically* processes information during these extreme events. Make no mistake: during cascading failures, the network is in fact *computing* its new stable state (attractor), so understanding this computation can help understand the dynamics here.

While the aforementioned quantitative studies of computation on networks are interesting, they do not directly measure the information dynamics claimed; e.g. none of the purported measures of information transfer properly measure directed, dynamic flows of information. Measures of model or task specific properties (e.g. in [61, 67, 80, 130]) are qualitatively appealing but give no insights into the underlying quantitative nature of the information dynamics, while mutual information between random pairs of nodes (by Ribeiro et al. [132] and Solé and Valverde [67]) measures dynamic correlation across the collective which may result from an information transfer but is not a measure of it. We also note the more generic measures of "information transfer" and "efficiency in transporting information" presented in [69, 124] respectively, however they are static measures of structure rather than measures of a directed, dynamic flow of information.[16]

A framework for the information dynamics of distributed computation would allow significant insights into computation in networks. In particular, it could be used to produce satisfactory quantitative insights into whether the computational properties of networks are indeed maximised near order-chaos phase transitions. It could also clarify the relationship between damage spreading and information transfer. Furthermore, such a framework would be used to provide quantitative answers on how network structure gives rise to computational properties.

In this section, we introduce RBNs as an important model of time-series dynamics on networks. RBNs are illustrative of complex systems perspectives in this domain. A model for cascading failures on networks will be introduced in Sect. 6.2.1. These two models will be used in Chap. 6 to study the nature of information dynamics in networks and phase transitions.

2.4.1 Random Boolean Networks as a Model of Dynamic Network Behaviour

Random Boolean networks (RBNs) are a class of generic discrete dynamical network models. They are particularly important in artificial life, since they were proposed as models of gene regulatory networks by Kauffman [16]. Indeed Boolean networks have been successfully used to model various GRNs (e.g. the regulatory network controlling metabolism in *E. coli* [139, 140] and the cell-cycle regulatory network of fission yeast [141]). They are also useful as generalised models of computation in networks. See also [131] for another thorough introduction to RBNs.

---

[16] The work in [69] is part of an interesting trend towards the use of information-theoretic measures to study network topology, e.g. see also [70]. Though information-theoretic, since these measures study topology rather than time-series dynamics they remain out of scope here.

An RBN consists of N nodes in a directed *network* structure. The nodes take *Boolean* state values, and update their state values in time as a deterministic function of the state values of the nodes from which it has incoming links. The network topology (i.e. the adjacency matrix) is determined at *random*, subject to whether the in-degree for each node is constant or stochastically determined given an average in-degree $\overline{K}$ (giving a Poissonian distribution). It is also possible to bias the network structure, e.g. toward scale-free degree distribution [142]. Given the topology, the deterministic Boolean function or lookup table by which each node computes its next state from its neighbours is also decided at *random* for each node, subject to a probability p of producing "1" outputs (p close to 1 or 0 gives low activity, close to 0.5 gives high activity). The nodes here are heterogeneous agents: there is no spatial pattern to the network structure (indeed there is no inherent concept of locality from a global perspective), nor do the nodes have the same update functions.[17] Importantly, the network structure and update functions for each node are held static in time ("quenched"). In classical RBNs (CRBNs), the nodes all update their states synchronously.[18]

It is worth noting that CAs are a sub-class of RBNs, with lattice-style ordering of the nodes and their connections, and homogeneous update rules [127]. So in a similar fashion to CAs, note that the synchronous nature of CRBNs, their Boolean states and deterministic update functions give rise to a global state $\mathbf{X}$ for the network, with deterministic transient trajectories through a state-space ultimately leading to either fixed or periodic attractors in finite-sized networks [127] (and see Sect. 2.3.2). Effectively, the transient is the period in which the network is *computing* its steady state attractor.

RBNs are known to exhibit three distinct phases of dynamics, depending on their parameters: ordered, chaotic and critical. The characteristics of these phases are similar to those described for CAs (see Sect. 2.3.2). At relatively low connectivity (i.e. low degree $\overline{K}$) or activity (i.e. p close to 0 or 1), the network is in an ordered phase, characterised by high stability of states and strong convergence of similar macro states in state space. Alternatively, at relatively high connectivity and activity, the network is in a chaotic phase, characterised by low stability of states and divergence of similar macro states. In the critical phase (described as the *edge of chaos* [15]), there is percolation in nodes remaining static or updating their values, and uncertainty in the convergence or divergence of similar macro states. The phase is also associated with large correlation lengths [103], similar to complex behaviour in CAs.

This phase transition is typically quantified using a measure of sensitivity to initial conditions, or damage spreading. We will describe a measure for this purpose in Sect. 6.1.1. It is important to note that (as per all many-body phase transitions, e.g. in

[17] Though, of course either of these can arise at random.

[18] There has been some debate about the best updating scheme to model GRNs [143], and variations on the synchronous CRBN model are known to produce different behaviours. However, the relevant phase transitions are known to exist in all updating schemes, and their properties depend more on the network size than on the updating scheme [144]. As such, the use of CRBNs is justified for ensemble studies such as ours [128].

the Ising model [145]) RBNs only exhibit a phase transition with truly discontinuous changes in system properties with continuous changes in parameters in the infinite-size limit [132]. Finite-size networks exhibit *approximate* phase transitions with: a continuous change in system properties with respect to parameters [132], variation in these properties over network realisations (which becomes lower at larger network sizes) [61, 132], and a shift of the critical point to a region centred on larger connectivities [144].

Much has been speculated on the possibility that gene regulatory and other biological networks function in (or evolve to) the critical regime (see [131]). It has been suggested that computation occurs more naturally with the balance of order and chaos there [15], possibly with information storage, propagation and processing capabilities maximised [16]. This is of course generalised in the "edge of chaos" hypothesis [15]: that systems exhibiting critical dynamics in the vicinity of a phase transition maximise their computational capability (see earlier discussion in Sect. 2.3.3). Here we seek to improve on previous attempts to measure these computational properties, with a thorough quantitative study of the information dynamics in RBNs.

2.5 Guided Self-Organisation

The principle of self-organisation is well known to offer the advantages of flexibility, robustness and scalability over centralised systems [73]. As such, it is often employed in the design of artificial systems for which these traits are desirable. Most self-organised solutions are currently designed in an ad-hoc manner, since the fundamental nature of self-organisation remains poorly understood. Some designers use an approach specifically tailored to the problem, e.g. [34]. More generally, designers use a genetic algorithm (GA) or programming (GP) approach, with fitness functions measuring specific achievement of the task required of the system (*task-based evolution*), e.g. [146].

Task-based evolution, this incumbent method of designing self-organised systems, can be impractical. Hand-crafting fitness functions for every task can be time-consuming and tedious, and requires specialised human understanding of the task. It has the potential to under-specify the problem (thereby solving a different task) or perhaps over-specify it (leading to an inflexible design). Also, the intelligent designer may not be completely sure of how to measure performance of the required task, or this may be difficult (e.g. measuring speed may require extra sensors). Coupling evolution with learning can be helpful, however specifying rewards for reinforcement learning suffer the same problem regarding measurement of the task [147]. Furthermore, if the initial task-based fitness landscape is flat and features no gradients, task-based evolution has no foothold around which to begin designing a solution. Finally, evolution often delivers intricate solutions of which (human) system managers cannot understand the inner workings: this is particularly undesirable for critical systems where maintenance or prediction of behaviour is required.

As an alternative, the concept of *guided self-organisation* proposes the use of information-theoretic measures to guide the emergence of required information processing structure in self-organised systems [10]. This perspective has been prompted by observations of complexity to grow or necessary information-theoretic structure to emerge during task-based evolution. For example, growth of complexity has been observed during evolution in *artificial life* (ALife) simulation environments, e.g.: by measuring "physical complexity" in Avida [31], and under certain conditions by measuring neural (TSE) complexity of evolved agents in Poly-World [148–150]. Looking at evolution for particular tasks, Prokopenko et al. [151] observed coordination (measured as excess entropy, see Eq. (2.18)) to increase in snake-like robots evolved for maximum velocity, and [152] observed a decrease in entropy in a swarm evolved for coordinated motion. Also, Ay et al. [153] observe that predictive information (see Eq. (2.20) with $k = 1$) is maximised for an autonomous robot exhibiting behaviour that is both "explorative and sensitive to the environment".

These observations suggest that such information-theoretic measures could be used themselves to guide self-organisation. This idea is fundamentally based on the theory that information structure is vital to the emergence of self-organised intelligence [154]. The concept could provide a consistent framework for the evolutionary design of self-organised systems, using template-based evolution for fundamental computational tasks that underpin the system goal. In theory, this approach would be able to produce useful structure where task-based evolution faces initially flat task-based fitness landscapes, perhaps serving as a platform from which to launch better-equipped task-based evolution. Furthermore, it may provide solutions which are simpler for humans to understand in terms of the underlying information dynamics. Perhaps most important is the potential for this approach to provide insight into the emergence rather than engineering of intelligence [154], and thereby facilitate unsupervised learning.

Several examples of successful guided self-organisation using information-theoretic measures exist in the literature. An interesting example is the concept of *empowerment* presented by Klyubin et al. [155, 156]. This concept refers to an agent's self-perception of its influence over the environment and is measured as the channel capacity of an agent's perception-action loop. Maximisation of empowerment has been shown to induce a necessary structure in an agent's behaviour, and indeed such maximisation has been suggested to be an intrinsic selection pressure.[19] Sporns and Lungarella [157] have evolved hand-eye co-ordination to grab a moving object using maximisation of neural (TSE) complexity. Interestingly, they demonstrated that this solution contained more intrinsic diversity than solutions from task-driven evolution; the increased diversity may afford greater flexibility to the system. Prokopenko et al. [73] were able to evolve fast-moving snake-like robots using maximisation of the excess entropy (see Eq. (2.18)) as an information-theoretic measure of co-ordination.

---

[19] The justification or otherwise of the suggestion that *natural* evolution is driven by the intrinsic forces of information processing is irrelevant to whether information-driven design can be used as a successful tool for *artificial* systems.

Also, Sperati et al. [158] have observed interesting periodic behaviour and complex structure in groups of robots which were evolved to maximise their mutual information. Note that the "guiding" could be through learning rather than evolution, e.g. the concept of "homeokinesis" [147, 159] which seeks to develop information structure in an agent by having it adapt to minimise the error of an internal model of its behaviour.

We suggest that utilising an understanding of distributed computation is a key approach for guided self-organisation using information-theoretical measures. Fernández and Solé [103] observe that since biological systems perform computations, there is an evolutionary pay-off in nature for this capability. There is good reason to expect this in artificial systems too, since any task we wish the system to achieve involves some form of computation. Indeed, Von Neumann believed that an understanding of distributed computation in CAs "would be essential …for *designing* artificial complex systems" ([88] describing [96]). To be specific, we suggest that the *information dynamics* of distributed computation provide the most intuitive basis here. As previously discussed, these information dynamics are the primitive functions of Turing universal computation, i.e. *information storage*, *transfer* and *modification*. As such, using a framework for distributed computation allows us to *target the design* toward the computational requirements of the task at hand, i.e. selecting either the most relevant computational function as the fitness function, or balancing the functions in a more sophisticated manner. Indeed, this perspective will provide insights to the related field of *unconventional computation*, which specifically seeks to design systems with greater capabilities than traditional computers [160]. Importantly, using such a framework provides a basis through which to understand the computation carried out by the solution. Also, guiding a system toward the building blocks of distributed computation is an intuitive way to facilitate the emergence of collective intelligence.

Information transfer is an important candidate fitness function to be investigated here. It has been conjectured that information transfer can give rise to interesting behaviour and induce necessary structure in a multi-agent system [73]. One inspiration of this viewpoint is the aforementioned concept of empowerment [155, 156], which has been shown to induce necessary structure in an agent's behaviour. This concept is relevant here since it alludes to information transfer in being quantified as the channel capacity between an agent's actuators and sensors through the environment. Also, several authors have attributed a key role to information transfer in facilitating the emergence of complex computation near the critical dynamics of order-chaos phase transitions [15, 17, 67]. Some have inferred maximisation of information transfer in the critical state [17, 67], however evidence has not yet been provided from a directed, dynamic measure of information transfer. Also, information transfer is said to be manifested in "information cascades" spreading across schools of fish [75]. Such mechanisms are biologically crucial, because they can transmit information at a faster speed than that of an incoming predator, with such computational capability providing an evolutionary advantage. We will investigate the manner in which information transfer can be harnessed in approaches to guided self-organisation in Sect. 8.3.

2.6 Opportunity to Quantify the Information Dynamics of Distributed Computation

In this chapter we have described complex systems, and the growing extent to which the perspective of distributed computation (using measures of information theory) is being used for the analysis and design of these systems. We have focussed on three specific domains (cellular automata, network dynamics and guided self-organisation) in order to describe current understanding of distributed computation. Each of these cases highlighted the specific need for and opportunity to provide a framework to quantify the information dynamics of distributed computation in complex systems. In particular, they have highlighted the need for these operations to be quantified on a local scale in space and time. We hypothesise that if we can describe and quantify distributed computation in terms of information storage, transfer and modification, then we will be better able to understand distributed computation in nature and its sources of complexity.

For CAs in particular, such a framework has the potential to quantitatively confirm the computational role of emergent structures, making a fundamental contribution to our understanding of the nature of distributed computation. As such, CAs will be used as the primary application area to study the measures we propose for this framework in the coming chapters. We emphasise that our approach is not to quantify computation or overall complexity, nor to identify universal computation or determine what is being computed. It is simply intended to quantify the component operations in space-time. We will examine how these operations interact in order to give rise to complex computation.

Similarly, we will investigate whether these information dynamics are maximised in order-chaos phase transitions. Phase transitions in networks provide a particularly important application area for these investigations. Furthermore, this framework has the potential to provide widely-anticipated insights into the dynamics in networks, and how these relate to topology, in a manner that can be compared across network types and address concepts of general interest (e.g. information transfer).

Crucially, considerations of information storage [72–74], transfer [75, 76, 161] and modification [79, 80, 83] extend well beyond these central application areas. We believe that the development of such a framework, and the insights it would provide from these application areas, would have significant impact on the whole of complex systems science. Importantly, it would be relevant in both analysis (e.g. in computational neuroscience as we demonstrate in Sect. 8.2), and design (i.e. in the context of guided self-organisation as we show in Sect. 8.3). In the next chapter we will begin describing our original contribution of this framework by considering the operation of information storage.

References

1. M. Prokopenko, F. Boschietti, A.J. Ryan, An information-theoretic primer on complexity, self-organization, and emergence. Complexity **15**(1), 11–28 (2009)
2. M. Mitchell, *Complexity: a guided tour* (Oxford University Press, New York, 2009)
3. P. Gong, C. van Leeuwen, Distributed dynamical computation in neural circuits with propagating coherent activity patterns. PLoS Comput. Biol. **5**(12), e1000611 (2009)
4. D. Peak, J.D. West, S.M. Messinger, K.A. Mott, Evidence for complex, collective dynamics and emergent, distributed computation in plants. Proc. Nat. Acad. Sci. U.S.A. **101**(4), 918–922 (2004)
5. D.J. Watts, S. Strogatz, Collective dynamics of 'small-world' networks. Nature **393**, 440–442 (1998)
6. A.-L. Barabási, R. Albert, Emergence of scaling in random networks. Science **286**(5439), 509–512 (1999)
7. D.J. Watts, *Six Degrees: the Science of a Connected Age* (Norton, New York, 2003)
8. A.-L. Barabási, Scale-free networks: decade and beyond. Science **325**(5939), 412–413 (2009)
9. M. Mitchell, Complex systems: network thinking. Artif. Intell. **170**(18), 1194–1212 (2006)
10. M. Prokopenko, Guided self-organization. HFSP J. **3**(5), 287–289 (2009)
11. F. Heylighen, P. Cilliers, C. Gershenson, Complexity and philosophy, in Complexity, Science and Society, ed. by J. Bogg, R. Geyer (Radcliffe Publishing, Oxford, 2007).
12. C.R. Shalizi, Methods and techniques in complex systems science: an overview, in *Complex Systems Science in Biomedicine*, ed. by T.S. Deisboeck, J.Y. Kresh (Springer, Berlin, 2006), pp. 33–114
13. J. Gleick, *Chaos: Making a New Science* (Penguin, New York, 1988)
14. J.R. Gribbin, *Deep Simplicity: Chaos, Complexity and the Emergence of Life* (Penguin, London, 2005)
15. C.G. Langton, Computation at the edge of chaos: phase transitions and emergent computation. Physica D **42**(1–3), 12–37 (1990)
16. S.A. Kauffman, *The Origins of Order: Self-Organization and Selection in Evolution* (Oxford University Press, New York, 1993)
17. O. Miramontes, Order-disorder transitions in the behavior of ant societies. Complexity **1**(3), 56–60 (1995)
18. M. Mitchell, J.P. Crutchfield, P.T. Hraber, Dynamics, computation, and the "edge of chaos": a re-examination, in *Complexity: Metaphors, Models, and Reality, ser. Santa Fe Institute Studies in the Sciences of Complexity*, ed. by G. Cowan, D. Pines, D. Melzner, vol. 19, (Addison-Wesley, Reading, 1994), pp. 497–513
19. D.P. Feldman, C.S. McTague, J.P. Crutchfield, The organization of intrinsic computation: complexity-entropy diagrams and the diversity of natural information processing. Chaos **18**(4), 043106 (2008)
20. C.R. Shalizi, K.L. Shalizi, R. Haslinger, Quantifying self-organization with optimal predictors. Phys. Rev. Lett. **93**(11), 118701 (2004)
21. D. Polani, Foundations and formalizations of self-organization, in *Advances in Applied Self-organizing Systems, ser. Advanced Information and Knowledge Processing*, ed. by M. Prokopenko (Springer, London, 2008), pp. 19–37
22. L. Correia, Self-organisation: a case for embodiment, in *Proceedings of the Evolution of Complexity Workshop at Artificial Life X: the 10th International Conference on the Simulation and Synthesis of Living Systems* (MIT Press, Bloomington, 2006), pp. 111–116
23. C. Prehofer, C. Bettstetter, Self-organization in communication networks: principles and design paradigms. IEEE Commun. Mag. **43**(7), 78–85 (2005)
24. P. Bak, C. Tang, K. Wiesenfeld, Self-organized criticality: an explanation of the 1/f noise. Phys. Rev. Lett. **59**(4), 381–384 (1987)
25. M. Buchanan, *Ubiquity* (Phoenix, London, 2001)

26. J. Léveillé, M. Versace, S. Grossberg, Running as fast as it can: how spiking dynamics form object groupings in the laminar circuits of visual cortex. J. Comput. Neurosci. **28**(2), 323–346 (2010)
27. E. Bonabeau, M. Dorigo, G. Theraulaz, *Swarm Intelligence: from Natural to Artificial Systems* (Oxford University Press, New York, 1999)
28. F. Ratnieks, Outsmarted by ants. Science **436**, 465 (2005)
29. S. Strogatz, *Sync: the Emerging Science of Spontaneous Order* (Hyperion Books, New York, 2003)
30. R. Pastor-Satorras, A. Vespignani, *Evolution and Structure of the Internet: a Statistical Physics Approach* (Cambridge University Press, Cambridge, 2004)
31. C. Adami, What is complexity? BioEssays **24**(12), 1085–1094 (2002)
32. M. Prokopenko, P. Wang, P. Valencia, D. Price, M. Foreman, A. Farmer, Self-organizing hierarchies in sensor and communication networks. Artif. Life **11**(4), 407–426 (2005)
33. M. Prokopenko, P. Wang, D. Price, Complexity metrics for self-monitoring impact sensing networks, in *Proceedings of the, NASA/DoD Conference on Evolvable Hardware (EH'05)* (IEEE Computer Society, Washington, 2005), pp. 239–246
34. J. Wessnitzer, A. Adamatzky, C. Melhuish, Towards self-organized robot formations: a decentralised approach, in *Proceedings of the Towards Intelligent Mobile Robots* (TIMR) Conference, Manchester, 2001)
35. M. Gell-Mann, *The Quark and the Jaguar* (W.H. Freeman, New York, 1994)
36. C.E. Shannon, A mathematical theory of communication. Bell Syst. Tech. J. **27**(379–423), 623–656 (1948)
37. T.M. Cover, J.A. Thomas, *Elements of Information Theory* (Wiley, New York, 1991)
38. D.J. MacKay, *Information Theory, Inference, and Learning Algorithms* (Cambridge University Press, Cambridge, 2003)
39. J.P. Crutchfield, D.P. Feldman, Regularities unseen, randomness observed: levels of entropy convergence. Chaos **13**(1), 25–54 (2003)
40. G. Tononi, O. Sporns, G. Edelman, A measure for brain complexity: relating functional segregation and integration in the nervous system. Proc. Nat. Acad. Sci. **91**(11), 5033–5037 (1994)
41. A.S. Klyubin, D. Polani, C.L. Nehaniv, Empowerment: a universal agent-centric measure of control, in *Proceedings of the IEEE Congress on Evolutionary Computation, Edinburgh, Scotland*, vol. 1 (IEEE Press, Edinburgh, 2005), pp. 128–135
42. P. Grassberger, Toward a quantitative theory of self-generated complexity. Int. J. Theor. Phys. **25**(9), 907–938 (1986)
43. W. Bialek, I. Nemenman, N. Tishby, Complexity through nonextensivity. Physica A **302**(1–4), 89–99 (2001)
44. C.D. Manning, H. Schütze, *Foundations of Statistical Natural Language Processing* (The MIT Press, Cambridge, 1999)
45. J. Dasan, T.R. Ramamohan, A. Singh, P.R. Nott, Stress fluctuations in sheared Stokesian suspensions. Phys. Rev. E **66**(2), 021409 (2002)
46. T. Schreiber, Interdisciplinary application of nonlinear time series methods–the generalized dimensions. Phys. Rep. **308**, 1–64 (1999)
47. H. Kantz, T. Schreiber, *Nonlinear Time Series Analysis* (Cambridge University Press, Cambridge, 1997)
48. M.R. DeWeese, M. Meister, How to measure the information gained from one symbol, Network: Comp. Neural Syst. **10**, 325–340 (1999)
49. P.F. Verdes, Assessing causality from multivariate time series, Phys. Rev. E, 72(2), 026 222, (2005).
50. C.R. Shalizi, *Causal architecture, complexity and self-organization in time series and cellular automata* (University of Wisconsin-Madison, Ph.D. Dissertation, 2001)
51. C.R. Shalizi, R. Haslinger, J.-B. Rouquier, K.L. Klinkner, C. Moore, Automatic filters for the detection of coherent structure in spatiotemporal systems. Phys. Rev. E **73**(3), 036104 (2006)

52. T. Helvik, K. Lindgren, M.G. Nordahl, Local information in one-dimensional cellular automata, in *Proceedings of the International Conference on Cellular Automata for Research and Industry, Amsterdam*, ed. by P.M. Sloot, B. Chopard, A.G. Hoekstra, ser. Lecture Notes in Computer Science, vol. 3305, (Springer, Berlin, 2004), pp. 121–130
53. T. Schreiber, Measuring information transfer. Phys. Rev. Lett. **85**(2), 461–464 (2000)
54. J. Pahle, A.K. Green, C.J. Dixon, U. Kummer, Information transfer in signaling pathways: a study using coupled simulated and experimental data. BMC Bioinform. **9**, 139 (2008)
55. A. Kraskov, H. Stögbauer, P. Grassberger, Estimating mutual information. Phys. Rev. E **69**(6), 066138 (2004)
56. A. Kraskov, Synchronization and interdependence measures and their applications to the electroencephalogram of epilepsy patients and clustering of data, ser. Publication Series of the John von Neumann Institute for Computing. Ph.D. thesis, vol. 24, John von Neumann Institute for Computing, Jülich, 2004.
57. L. Kozachenko, N. Leonenko, A statistical estimate for the entropy of a random vector. Prob. Inf. Transm. **23**, 9–16 (1987)
58. D. Polani, Information: currency of life? HFSP J. **3**(5), 307–316 (2009)
59. J.J. Gibson, *The Ecological Approach to Visual Perception* (Houghton Mifflin Company, Boston, 1979)
60. J.P. Crutchfield, The calculi of emergence: computation, dynamics and induction. Physica D **75**(1–3), 11–54 (1994)
61. P. Rämö, S. Kauffman, J. Kesseli, O. Yli-Harja, Measures for information propagation in Boolean networks. Physica D **227**(1), 100–104 (2007)
62. A. Wuensche, Classifying cellular automata automatically: finding gliders, filtering, and relating space-time patterns, attractor basins, and the Z parameter. Complexity **4**(3), 47–66 (1999)
63. N. Ay, E. Olbrich, N. Bertschinger, J. Jost, A unifying framework for complexity measures of finite systems, in *Proceedings of the European Conference on Complex Systems (ECCS '06), Oxford, UK*, ed. by J. Jost, F. Reed-Tsochas, P. Schuster, Santa Fe Institute Working Paper 06–08-028. 2006, p. 80.
64. E. Olbrich, N. Bertschinger, N. Ay, J. Jost, How should complexity scale with system size? Eur. Phys. J. B **63**(3), 407–415 (2008)
65. J.P. Crutchfield, K. Young, Inferring statistical complexity. Phys. Rev. Lett. **63**(2), 105 (1989)
66. C.R. Shalizi, J.P. Crutchfield, Computational mechanics: pattern and prediction, structure and simplicity. J. Stat. Phys. **104**, 817–879 (2001)
67. R.V. Solé, S. Valverde, Information transfer and phase transitions in a model of internet traffic. Physica A **289**(3–4), 595–605 (2001)
68. M. Mitchell, P.T. Hraber, J.P. Crutchfield, Revisiting the edge of chaos: evolving cellular automata to perform computations. Complex Syst. **7**, 89–130 (1993)
69. R.V. Solé, S. Valverde, Information theory of complex networks: on evolution and architectural constraints, in *Complex Networks*, ed. by E. Ben-Naim, H. Frauenfelder, Z. Toroczkai, ser. Lecture Notes in Physics, vol. 650, (Springer, Berlin, 2004), pp. 189–207
70. M. Piraveenan, M. Prokopenko, A.Y. Zomaya, Assortativeness and information in scale-free networks. Eur. Phys. J. B **67**(3), 291–300 (2009)
71. M.G. Kitzbichler, M.L. Smith, S.R. Christensen, E. Bullmore, Broadband criticality of human brain network synchronization. PLoS Comput. Biol. **5**(3), e1000314 (2009)
72. R. Morgado, M. Ciesla, L. Longa, F.A. Oliveira, Synchronization in the presence of memory. Europhys. Lett. **79**(1), 10002 (2007)
73. M. Prokopenko, V. Gerasimov, I. Tanev, Evolving spatiotemporal coordination in a modular robotic system, in *Proceedings of the Ninth International Conference on the Simulation of Adaptive Behavior (SAB'06), Rome*, ed. by S. Nolfi, G. Baldassarre, R. Calabretta, J. Hallam, D. Marocco, J.-A. Meyer, D. Parisi, ser. Lecture Notes in Artificial Intelligence, vol. 4095, (Springer, Berlin, 2006), pp. 548–559
74. K.I. Goh, A.L. Barabási, Burstiness and memory in complex systems. Europhys. Lett. **81**(4), 48002 (2008)

75. I. Couzin, R. James, D. Croft, J. Krause, Social organization and information transfer in schooling fishes, in Fish Cognition and Behavior, ed. by B.C.K. Laland, J. Krause, ser. Fish and Aquatic Resources(Blackwell Publishing, New York, 2006), pp. 166–185.
76. J.A. Brown, J.A. Tuszynski, A review of the ferroelectric model of microtubules. Ferroelectrics **220**, 141–156 (1999)
77. M.H. Jakubowski, K. Steiglitz, R. Squier, Information transfer between solitary waves in the saturable Schrödinger equation. Phys. Rev. E **56**(6), 7267 (1997)
78. A. Adamatzky (ed.), *Collision-Based Computing* (Springer, Berlin, 2002)
79. D.E. Edmundson, R.H. Enns, Fully 3-dimensional collisions of bistable light bullets. Opt. Lett. **18**, 1609–1611 (1993)
80. O. Kinouchi, M. Copelli, Optimal dynamical range of excitable networks at criticality. Nat. Phys. **2**(5), 348–351 (2006)
81. J.J. Atick, Could information theory provide an ecological theory of sensory processing? Netw. Comput. Neural Syst. **3**(2), 213 (1992)
82. M.A. Sánchez-Montañés, F.J. Corbacho, Towards a new information processing measure for neural computation, in *Proceedings of the International Conference on Artificial Neural Networks (ICANN 2002), Madrid, Spain*, ed. by J. Dorronsoro, ser. Lecture Notes in Computer Science, vol. 2415, (Springer, Berlin, 2002), pp. 637–642
83. T. Yamada, K. Aihara, Spatio-temporal complex dynamics and computation in chaotic neural networks, in *Proceedings of the IEEE Symposium on Emerging Technologies and Factory Automation (ETFA '94), Tokyo* (IEEE, New York, 1994), pp. 239–244
84. D. Coffey, Self-organization, complexity and chaos: the new biology for medicine. Nat. Med. **4**(8), 882 (1998)
85. T.Q. Tung, T. Ryu, K.H. Lee, D. Lee, Inferring gene regulatory networks from microarray time series data using transfer entropy, in *Proceedings of the Twentieth IEEE International Symposium on Computer-Based Medical Systems (CBMS '07), Maribor, Slovenia*, ed. by P. Kokol, V. Podgorelec, D. Mičetič-Turk, M. Zorman, M. Verlič (Los Alamitos, IEEE, 2007), pp. 383–388
86. K. Ishiguro, N. Otsu, M. Lungarella, Y. Kuniyoshi, Detecting direction of causal interactions between dynamically coupled signals. Phys. Rev. E **77**(2), 026216 (2008)
87. X.S. Liang, Information flow within stochastic dynamical systems. Phys. Rev. E **78**(3), 031113 (2008)
88. M. Mitchell, Computation in cellular automata: a selected review, in *Non-Standard Computation*, ed. by T. Gramss, S. Bornholdt, M. Gross, M. Mitchell, T. Pellizzari (VCH Verlagsgesellschaft, Weinheim, 1998), pp. 95–140
89. S. Wolfram, Universality and complexity in cellular automata. Physica D **10**(1–2), 1–35 (1984)
90. S. Wolfram, Cellular automata as models of complexity. Nature **311**(5985), 419–424 (1984)
91. S. Wolfram, Computation theory of cellular automata. Commun. Math. Phys. **96**(1), 15–57 (1984)
92. S. Wolfram, *A New Kind of Science* (Wolfram Media, Champaign, 2002)
93. M. Mitchell, J.P. Crutchfield, P.T. Hraber, Evolving cellular automata to perform computations: mechanisms and impediments. Physica D **75**, 361–391 (1994)
94. J.E. Hanson, J.P. Crutchfield, The attractor-basin portait of a cellular automaton. J. Stat. Phys. **66**, 1415–1462 (1992)
95. H. Gutowitz, C. Domain, The topological skeleton of cellular automaton dynamics. Physica D **103**(1–4), 155–168 (1997)
96. J. Von Neumann, Theory of self-reproducing automata ed. by A.W. Burks, (University of Illinois Press, Urbana, 1966).
97. J.T. Lizier, M. Prokopenko, A.Y. Zomaya, Information modification and particle collisions in distributed computation. Chaos **20**(3), 037109 (2010)
98. J.T. Lizier, M. Prokopenko, A.Y. Zomaya, Coherent information structure in complex computation, Theory Biosci. **131**(3), 193–203 (2012), doi:10.1007/s12064-011-0145-9.
99. L. Gray, A mathematician looks at wolfram's new kind of science. Am. Math. Soc. **50**(2), 200–211 (2003)

100. J.-B. Rouquier, Cimula–a cellular automata analyser (2005), Université de Lyon, Software. http://cimula.sourceforge.net/
101. S. Lloyd, *Programming the Universe* (Vintage Books, New York, 2006)
102. J.L. Casti, Chaos, Gödel and truth, in *Beyond belief: randomness, prediction and explanation in science*, ed. by J.L. Casti, A. Karlqvist (CRC Press, Boca Raton, 1991), pp. 280–327
103. P. Fernández, R.V. Solé, The role of computation in complex regulatory networks, in *Scale-free Networks and Genome Biology*, ed. by E.V. Koonin, Y.I. Wolf, G.P. Karev (Georgetown, Landes Bioscience, 2006), pp. 206–225
104. J. H. Conway, What is life? in Winning ways for your mathematical plays, ed. by E. Berlekamp, J.H. Conway, R. Guy, vol. 2, ch. 25(Academic Press, New York, 1982), pp. 927–962.
105. M. Cook, Universality in elementary cellular automata. Complex Syst. **15**(1), 1–40 (2004)
106. K. Lindgren, M.G. Nordahl, Universal computation in simple one-dimensional cellular automata. Complex Syst. **4**, 299–318 (1990)
107. M.H. Jakubowski, K. Steiglitz, R.K. Squier, Computing with solitons: a review and prospectus. Multiple-Valued Logic **6**(5–6), 439–462 (2001)
108. D. Eppstein, Searching for spaceships, in *More Games of No Chance*, ed. by R.J. Nowakowski, ser. MSRI Publications, vol. 42, (Cambridge Univ. Press, Cambridge, 2002), pp. 433–453
109. M. Mitchell, J.P. Crutchfield, R. Das, Evolving cellular automata with genetic algorithms: a review of recent work, in *Proceedings of the First International Conference on Evolutionary Computation and Its Applications, Moscow*, ed. by E. D. Goodman, W. Punch, V. Uskov (Russian Academy of Sciences, Moscow, 1996).
110. N. Boccara, J. Nasser, M. Roger, Particlelike structures and their interactions in spatiotemporal patterns generated by one-dimensional deterministic cellular-automaton rules. Phys. Rev. A **44**(2), 866–875 (1991)
111. B. Martin, A group interpretation of particles generated by one dimensional cellular automaton, Wolfram's 54 rule. Int. J. Mod. Phys. C **11**(1), 101–123 (2000)
112. G.J. Martinez, A. Adamatzky, H.V. McIntosh, Phenomenology of glider collisions in cellular automaton rule 54 and associated logical gates. Chaos, Solitons Fractals **28**(1), 100–111 (2006)
113. W. Hordijk, C.R. Shalizi, J.P. Crutchfield, Upper bound on the products of particle interactions in cellular automata. Physica D **154**(3–4), 240–258 (2001)
114. P. Grassberger, Long-range effects in an elementary cellular automaton. J. Stat. Phys. **45**(1–2), 27–39 (1986)
115. H.V. McIntosh, *Linear Cellular Automata* (Universidad Autónoma de Puebla, Puebla, 1990)
116. R. Badii, A. Politi, Thermodynamics and complexity of cellular automata. Phys. Rev. Lett. **78**(3), 444 (1997)
117. A. Lafusa, T. Bossomaier, Hyperplane localisation of self-replicating and other complex cellular automata rules, in *Proceedings of the, IEEE Congress on Evolutionary Computation, Edinburgh*, vol. 1 (IEEE Press, New York, 2005), pp. 844–849
118. P. Grassberger, New mechanism for deterministic diffusion. Phys. Rev. A **28**(6), 3666 (1983)
119. P. Grassberger, Information content and predictability of lumped and distributed dynamical systems. Physica Scripta **40**(3), 346 (1989)
120. J.E. Hanson, J.P. Crutchfield, Computational mechanics of cellular automata: an example. Physica D **103**(1–4), 169–189 (1997)
121. S.N. Dorogovstev, J.F.F. Mendes, *Evolution of networks: From biological nets to the Internet and WWW* (Oxford University Press, New York, 2005)
122. A.-L. Barabási, R. Albert, H. Jeong, Scale-free characteristics of random networks: the topology of the world-wide web. Physica A **281**, 69–77 (2000)
123. A.-L. Barabási, E. Bonabeau, Scale-free networks. Sci. Am. **288**, 50–59 (2003)
124. V. Latora, M. Marchiori, Efficient behavior of small-world networks. Phys. Rev. Lett. **87**(19), 198701 (2001)
125. F. Schweitzer, G. Fagiolo, D. Sornette, F. Vega-Redondo, A. Vespignani, D.R. White, Economic networks: the new challenges. Science **325**(5939), 422–425 (2009)

126. B. Derrida, Y. Pomeau, Random networks of automata: a simple annealed approximation. Europhys. Lett. **1**(2), 45–49 (1986)
127. A. Wuensche, *Attractor basins of discrete networks* (The University of Sussex, Ph.D. Dissertation, 1997)
128. C. Gershenson, Updating schemes in random Boolean networks: do they really matter?, in *Proceedings of the Ninth International Conference on the Simulation and Synthesis of Living Systems (ALife IX), Boston, USA*, ed. by J. Pollack, M. Bedau, P. Husbands, T. Ikegami, R.A. Watson (MIT Press, Cambridge, 2004), pp. 238–243
129. T. Rohlf, N. Gulbahce, C. Teuscher, Damage spreading and criticality in finite random dynamical networks. Phys. Rev. Lett. **99**(24), 248701 (2007)
130. E. Estrada, Information mobility in complex networks. Phys. Rev. E **80**(2), 026104 (2009)
131. C. Gershenson, Introduction to random Boolean networks, in *Proceedings of the Workshops and Tutorials of the Ninth International Conference on the Simulation and Synthesis of Living Systems (ALife IX), Boston, USA*, ed. by M. Bedau, P. Husbands, T. Hutton, S. Kumar, H. Suzuki, 2004, pp. 160–173.
132. A.S. Ribeiro, S.A. Kauffman, J. Lloyd-Price, B. Samuelsson, J.E.S. Socolar, Mutual information in random Boolean models of regulatory networks. Phys. Rev. E **77**(1), 011901 (2008)
133. A.S. Ribeiro, R.A. Este, J. Lloyd-Price, S.A. Kauffman, Measuring information propagation and retention in boolean networks and its implications to a model of human organizations. WSEAS Trans. Syst. **5**, 2935 (2006)
134. P. Crucitti, V. Latora, M. Marchiori, Model for cascading failures in complex networks. Phys. Rev. E **69**(4), 045104 (2004)
135. A.E. Motter, Y.-C. Lai, Cascade-based attacks on complex networks. Phys. Rev. E **66**(6), 065102 (2002)
136. J. Glanz, R. Perez-Pena, 90 seconds that left tens of millions of people in the dark, New York Times, August 26 2003.
137. C. Oosawa, M.A. Savageau, Effects of alternative connectivity on behavior of randomly constructed Boolean networks. Physica D **170**(2), 143–161 (2002)
138. Q. Lu, C. Teuscher, Damage spreading in spatial and small-world random Boolean networks (2009), arXiv:0904.4052. http://arxiv.org/abs/0904.4052
139. A. Samal, S. Jain, The regulatory network of e. coli metabolism as a Boolean dynamical system exhibits both homeostasis and flexibility of response. BMC Syst. Biol. **2**(1), 21 (2008)
140. M.W. Covert, E.M. Knight, J.L. Reed, M.J. Herrgard, B.O. Palsson, Integrating high-throughput and computational data elucidates bacterial networks. Nature **429**(6987), 92–96 (2004)
141. M.I. Davidich, S. Bornholdt, Boolean network model predicts cell cycle sequence of fission yeast. PLoS ONE **3**(2), e1672 (2008)
142. M. Aldana, Boolean dynamics of networks with scale-free topology. Physica D **185**(1), 45–66 (2003)
143. C. Darabos, M. Giacobini, M. Tomassini, Semi-synchronous activation in scale-free Boolean networks, in *Proceedings of the 9th European Conference on Artificial Life (ECAL), Lisbon, Portugal* ed. by F. Almeida e Costa, L. M. Rocha, E. Costa, I. Harvey, A. Coutinho, ser. Lecture Notes in Artificial Intelligence, vol. 4648 (Springer, Berlin, 2007), pp. 976–985.
144. C. Gershenson, Phase transitions in random Boolean networks with different updating schemes (2004), arXiv:nlin/0311008v1. http://arxiv.org/abs/nlin/0311008
145. L. Onsager, Crystal statistics. I. A two-dimensional model with an order-disorder transition. Phys. Rev. **65**(3–4), 117–149 (1944)
146. I. Tanev, T. Ray, A. Buller, Automated evolutionary design, robustness, and adaptation of sidewinding locomotion of a simulated snake-like robot. IEEE Trans. Robot **21**(4), 632–645 (2005)
147. R. Der, U. Steinmetz, F. Pasemann, Homeokinesis–a new principle to back up evolution with learning, in Computational Intelligence for Modelling, Control, and Automation, ser. Concurrent Systems Engineering Series, vol. 55 (IOS Press, Amsterdam, 1999), pp. 43–47.

148. L. Yaeger, O. Sporns, Evolution of neural structure and complexity in a computational ecology, in *Proceedings of the Tenth International Conference on Simulation and Synthesis of Living Systems (ALifeX), Bloomington, Indiana, USA*, ed. by L.M. Rocha, L.S. Yaeger, M.A. Bedau, D. Floreano, R.L. Goldstone, A. Vespignani (MIT Press, Cambridge, 2006), pp. 330–336
149. L. Yaeger, V. Griffith, O. Sporns, Passive and driven trends in the evolution of complexity, in *Proceedings of the Eleventh International Conference on the Simulation and Synthesis of Living Systems (ALifeXI), Winchester, UK*, ed. by S. Bullock, J. Noble, R. Watson, M.A. Bedau (MIT Press, Cambridge, 2008), pp. 725–732
150. L.S. Yaeger, How evolution guides complexity. HFSP J. **3**(5), 328–339 (2009)
151. M. Prokopenko, V. Gerasimov, I. Tanev, Measuring spatiotemporal coordination in a modular robotic system, in *Proceedings of the 10th International Conference on the Simulation and Synthesis of Living Systems (ALifeX), Bloomington, Indiana, USA*, ed. by L.M. Rocha, L.S. Yaeger, M.A. Bedau, D. Floreano, R.L. Goldstone, A. Vespignani (MIT Press, Cambridge, 2006), pp. 185–191
152. G. Baldassare, D. Parisi, S. Nolfi, Measuring coordination as entropy decrease in groups on linked simulated robots, in *Proceedings of the International Conference on Complex Systems (ICCS2004)*, (Boston, 2004) (to be published).
153. N. Ay, N. Bertschinger, R. Der, F. Güttler, E. Olbrich, Predictive information and explorative behavior of autonomous robots. Eur. Phys. J. B **63**(3), 329–339 (2008)
154. D. Polani, O. Sporns, M. Lungarella, How information and embodiment shape intelligent information processing, in *Proceedings of the 50th Anniversary Summit of Artificial Intelligence, New York* ed. by M. Lungarella, F. Iida, J. Bongard, R. Pfeifer, ser. Lecture Notes in Computer Science, vol. 4850 (Springer, Berlin, 2007), pp. 99–111.
155. A.S. Klyubin, D. Polani, C.L. Nehaniv, Keep your options open: an information-based driving principle for sensorimotor systems. PLoS ONE **3**(12), e4018 (2008)
156. A.S. Klyubin, D. Polani, C.L. Nehaniv, All else being equal be empowered, in *Proceedings of the 8th European Conference on Artificial Life (ECAL), Kent, UK* ed. by M.S. Capcarrere, A. A. Freitas, P.J. Bentley, C.G. Johnson, J. Timmis, ser. Lecture Notes in Computer Science, vol. 3630 (Springer, Berlin, 2005), pp. 744–753.
157. O. Sporns, M. Lungarella, Evolving coordinated behavior by maximizing information structure, in *Proceedings of the Tenth International Conference on Simulation and Synthesis of Living Systems (ALifeX), Bloomington, Indiana, USA*, ed. by L.M. Rocha, L.S. Yaeger, M.A. Bedau, D. Floreano, R.L. Goldstone, A. Vespignani (MIT Press, Cambridge, 2006), pp. 323–329
158. V. Sperati, V. Trianni, S. Nolfi, Evolving coordinated group behaviours through maximisation of mean mutual information. Swarm Intell. **2**(2–4), 73–95 (2008)
159. F. Hesse, G. Martius, R. Der, J.M. Herrmann, A sensor-based learning algorithm for the self-organization of robot behavior. Algorithms **2**(1), 398–409 (2009)
160. C. Teuscher, I. Nemenman, F.J. Alexander, Novel computing paradigms: Quo vadis? Physica D, 237(9), v-viii (2008).
161. M. Lungarella, O. Sporns, Mapping information flow in sensorimotor networks. PLoS Comput. Biol. **2**(10), e144 (2006)

Chapter 3
Information Storage

Information storage is considered an important aspect of the dynamics of many natural and man-made processes, for example: in human brain networks [1] and artificial neural networks [2], synchronisation between coupled systems [3], coordinated motion in modular robots [4], and in the dynamics of inter-event distribution times [5]. The term is still often used rather loosely or in a qualitative sense however, and as yet we do not have a good understanding of how information storage interacts with information transfer and modification to give rise to distributed computation.

In this chapter we outline methods to quantify information storage in distributed computation.[1] We define the concept as *the information in an agent or variable's past that can be used to predict its future*. We describe how *total* storage is captured by the existing measure excess entropy in Sect. 3.1, and introduce active information storage in Sect. 3.2 to capture the amount of storage that is *currently in use*. In particular we focus on describing how these measures can be used to quantify information storage on a *local scale* in space-time in distributed computation (in Sects. 3.1.2 and 3.2.1). Our perspective of distributed computation is important, providing the perspective that information can not only be stored internally by an agent, but also stored in its environment for later retrieval.

We present the first application of local profiles of both measures to cellular automata in Sect. 3.3. As hypothesised in Sect. 2.3 these applications provide the first quantitative evidence that blinkers are the dominant information storage entities there. This result is significant in marrying these quantitative measures of information storage with the popularly-understood qualitative notion of its embodiment in distributed computation. The application also demonstrates the manner in which these two measures of information storage are distinct but complementary.

[1] The methods and results in this chapter were first reported in [6, 7].

J. T. Lizier, *The Local Information Dynamics of Distributed Computation in Complex Systems*, Springer Theses, DOI: 10.1007/978-3-642-32952-4_3,

© Springer-Verlag Berlin Heidelberg 2013

3.1 Excess Entropy as Total Information Storage

Discussion of information storage or memory in CAs has often focused on periodic structures (particularly in construction of universal Turing machines), e.g. [8]. However, information storage does not necessarily entail periodicity. The **excess entropy** (Eqs. 2.18, 2.19) more broadly encompasses all types of structure and memory by capturing correlations across all lengths of time, including non-linear effects. The predictive information formulation of the excess entropy in Eq. (2.19) (i.e. as the information from a system's past that can be used to predict its future) makes explicit that it is a measure of the total information storage in a system. In this section, we describe how the excess entropy is used to measure single-agent and collective information storage in Sect. 3.1.1. We also discuss how information storage in an agent's environment in a distributed computation increases its information storage capacity beyond its internal capability. subsequently in Sect. 3.1.2 we describe how the excess entropy can be localised in time and space.

3.1.1 Single-Agent and Collective Excess Entropy

We use the term *single-agent excess entropy* (or just excess entropy) to refer to measuring the quantity E_X for individual agents X or cells X_i using their one-dimensional time series of states. This is a measure of the *average* memory for *each agent.*

The predictive information form Eq. (2.20) $E_X(k) = I_{X^{(k)};X^{(k+)}}$, shows that the maximum excess entropy is the information capacity in sequences of k states $X^{(k)}$. In ECAs for example, this is 2^k bits. In the limit $k \to \infty$, this becomes infinite. Mathematically then, the information stored by a single agent can be larger than the information capacity of a single state. Where the agent takes direct causal influence from only a single past state (as in CAs), the meaning of its information storage being larger than its information capacity is not immediately obvious. For instance, a cell in an ECA could not store more than 1 bit of information in isolation. However, the cells in a CA are participating in a *distributed computation*: cyclic causal paths (facilitated by *bidirectional* links) effectively allow cells to store extra information in neighbours (even beyond the immediate neighbours), and to subsequently retrieve that information from those neighbours at a later point in time.While measurement of the excess entropy does not explicitly look for such *self-influence* communicated through neighbours, it is indeed the channel through which a significant portion of information can be communicated. This self-influence between semi-infinite past and future blocks being conveyed via neighbours is indicated by the curved arrows in Fig. 3.1a. It is akin to the use of stigmergy (indirect communication through the environment, e.g. see [9]) to communicate with oneself. Indeed, because information may be stored and retrieved from one's neighbours, an agent can store information regardless of whether it is causally connected with itself.

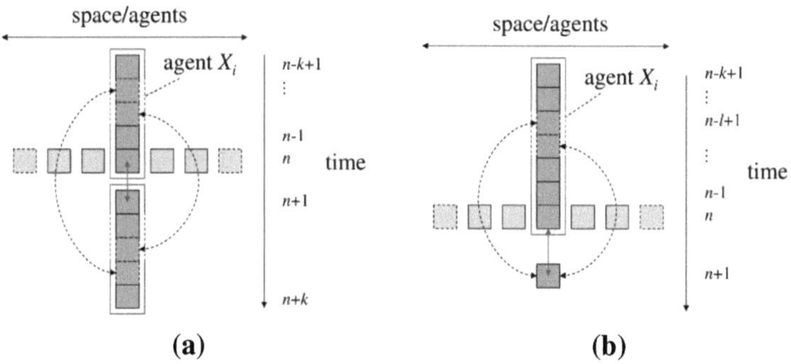

Fig. 3.1 Measures of single-agent information storage in distributed systems. **a** Excess entropy: *total* information from the cell's past that can be used to predict its future. **b** Active information storage: the information storage that is *currently in use* in determining the next state of the cell. The stored information can be conveyed directly through the cell itself or via neighbouring cells. (NB: This figure is reprinted from [7] with permission of Elsevier.)

Information storage exceeding single-state information capacity is then a perfectly valid result. Indeed in an infinite CA, each cell has access to an infinite number of neighbours in which to store an infinite amount of information that can later be used to influence its own future. Since the storage medium is *shared* by all cells though, one should not think about the total memory as the total number of cells N multiplied by this amount (i.e. to give NE_X).

The average total memory stored in a collective of agents (e.g. a set of neighbouring cells in a CA) is properly measured by the *collective excess entropy*. It is measured as *temporal* excess entropy of the agents using their two-dimensional time series of states. It is a joint temporal predictive information, i.e. the mutual information between the joint past $\mathbf{X}^{(k)}$ and future $\mathbf{X}^{(k^+)}$ of the agents:

$$E_{\mathbf{X}} = \lim_{k \to \infty} I_{\mathbf{X}^{(k)};\mathbf{X}^{(k^+)}}, \tag{3.1}$$

This collective measurement takes into account the inherent redundancy in the shared information storage medium (which NE_X does not). Collective excess entropy could be used for example to quantify the "undiscovered *collective memory* that may present in certain fish schools" [10].

As described in Appendix B, Grassberger found divergent collective excess entropy for several CA rules, including rule 22 [11, 12].[2] This infinite amount of collective memory implies a highly complex process, since in using strong long-range correlations a semi-infinite sequence "could store an infinite amount of information about its continuation" [13]. On the other hand, infinite collective excess entropy can

[2] Lindgren and Nordahl [13] also measured excess entropy (referred to as effective measure complexity) for some ECAs. They measured spatial excess entropies however, and we note that it is only temporal excess entropies which are interpretable as information storage from our perspective of distributed computation.

also be achieved for systems that only trivially utilise all of their available memory (see Appendix B). In attempting to quantify *local* information dynamics of distributed computation here, our focus is on information storage for *single agents or cells* rather than the joint information storage across the collective. Were the single-agent excess entropy found to be divergent (this has not been demonstrated), this may be more significant than for the collective case. This is because it would imply that all agents are *individually* strongly utilising the resources of the collective in a highly complex process.

Our focus is however on *locally* quantifying information storage in both *time* and space. We hypothesise this will provide much more detailed insights than single ensemble values into information storage structures and their involvement in distributed computation.

3.1.2 Local Excess Entropy

The **local excess entropy** is a measure of how much information a given agent is *currently* storing at a particular point in time. To derive it, note that (as per Sect. 2.2.2) the excess entropy of a process is actually the *expectation value* of the local excess entropy for the process at every time step [14].[3] Using the predictive information formulation,[4] the local excess entropy $e_X(n+1)$ of a process is simply the local mutual information between the semi-infinite past and future at the given time step $n+1$:

$$E_X = \langle e_X(n+1) \rangle_n , \tag{3.2}$$

$$e_X(n+1) = \lim_{k \to \infty} i(x_n^{(k)}; x_{n+1}^{(k^+)}), \tag{3.3}$$

$$e_X(n+1) = \lim_{k \to \infty} \log_2 \frac{p(x_n^{(k)}, x_{n+1}^{(k^+)})}{p(x_n^{(k)}) p(x_{n+1}^{(k^+)})}. \tag{3.4}$$

Recall from Sect. 2.2.1 that the semi-infinite past $x_n^{(k)}$ runs up until the previous time step n, while the semi-infinite future $x_{n+1}^{(k^+)}$ runs from the next time step $n+1$ onwards. The limit $k \to \infty$ is an important part of this definition [carried over from Eq. (2.19)], since correlations at all time scales should be included in the computation of information storage. Since this is not computationally feasible in general, we retain the following notation for finite-k estimates:

[3] This is as per Shalizi's original formulation of the local excess entropy in [14], however our presentation is for a single time-series rather than the light-cone formulation used there.

[4] Certainly the formulation of entropy rate overestimates in Eq. (2.18) could be used to directly form alternative localisations e'_X also, and in the limit $k \to \infty$ their averages E_X will be the same. However, only the local formulation from the predictive information captures the total information stored at a particular temporal point, which is our quantity of interest here.

$$E_X(k) = \langle e_X(n+1,k) \rangle_n , \tag{3.5}$$

$$e_X(n+1,k) = i(x_n^{(k)}; x_{n+1}^{(k^+)}), \tag{3.6}$$

$$= \log_2 \frac{p(x_n^{(k)}, x_{n+1}^{(k^+)})}{p(x_n^{(k)}) p(x_{n+1}^{(k^+)})}, \tag{3.7}$$

$$e_X(n+1) = \lim_{k \to \infty} e_X(n+1,k). \tag{3.8}$$

The notation is generalised for lattice systems (such as CAs) with *spatially-ordered* agents to represent the local excess entropy for cell X_i at time $n+1$ as:

$$e(i,n+1) = \lim_{k \to \infty} \log_2 \frac{p(x_{i,n}^{(k)}, x_{i,n+1}^{(k^+)})}{p(x_{i,n}^{(k)}) p(x_{i,n+1}^{(k^+)})}, \tag{3.9}$$

$$= \lim_{k \to \infty} e(i,n+1,k). \tag{3.10}$$

Local excess entropy is defined for every spatiotemporal point (i,n) in the system (where i is a spatial index and n is a time index). Note that the collective excess entropy $E_{\mathbf{X}}$ can also be localised, but only in time, to have $e_{\mathbf{X}}(n+1,k)$.

While the average excess entropy is always positive, the local excess entropy may in fact be positive or negative, meaning the past history of the cell can either positively inform us or actually *misinform* us about its future. An observer is misinformed where the semi-infinite past and future are relatively unlikely to be observed together as compared to their independent likelihoods. In other words, an observer is misinformed when the observed future is conditionally less likely given the observed past than without considering the past. In this situation we have $p(x_{n+1}^{(k^+)} \mid x_n^{(k)}) < p(x_{n+1}^{(k^+)})$ making the denominator of Eq. (3.4) greater than the numerator, and giving a negative value for $e_X(n+1)$.

3.2 Active Information Storage as Storage in Use

The excess entropy measures the total stored information which will be used *at some point* in the future of the state process of an agent. This information will possibly but not necessarily be used at the next time step $n+1$. Since the dynamics of computation unfold one step at a time, we are quite interested in how much of the stored information is actually *in use* at the next time step when the new process value is computed. As we will see in Chap. 5, this is particularly important in understanding how stored information interacts with information transfer in information processing. As such, we derive **active information storage** A_X as the average mutual information between the semi-infinite past of the process $X^{(k)}$ and its *next state* X', as opposed to its whole (semi-infinite) future:

$$A_X = \lim_{k\to\infty} A_X(k), \tag{3.11}$$

$$A_X(k) = I(X^{(k)}; X'). \tag{3.12}$$

We use $A_X(k)$ to represent finite-k estimates. The active information storage is represented in Fig. 3.1b. Of course, one could also define a collective active information storage $A_{\mathbf{X}}$.

3.2.1 Local Active Information Storage

Following Sect. 2.2.2, the **local active information storage** $a_X(n+1)$ is then a measure of the amount of information storage in use by the process at a particular time-step $n+1$. It is the local mutual information between the semi-infinite past of the process and its next state:

$$A_X = \langle a_X(n+1)\rangle_n, \tag{3.13}$$

$$a_X(n+1) = \lim_{k\to\infty} a_X(n+1,k), \tag{3.14}$$

$$A_X(k) = \langle a_X(n+1,k)\rangle_n, \tag{3.15}$$

$$a_X(n+1,k) = \lim_{k\to\infty} \log_2 \frac{p(x_n^{(k)}, x_{n+1})}{p(x_n^{(k)})p(x_{n+1})}, \tag{3.16}$$

$$= \lim_{k\to\infty} i(x_n^{(k)}; x_{n+1}). \tag{3.17}$$

As for the excess entropy, note that we have retained notation for finite-k estimates here.

Again, we generalise the measure for agent X_i in a lattice system as:

$$a(i,n+1) = \lim_{k\to\infty} \log_2 \frac{p(x_{i,n}^{(k)}, x_{i,n+1})}{p(x_{i,n}^{(k)})p(x_{i,n+1})}, \tag{3.18}$$

$$= \lim_{k\to\infty} a(i,n+1,k), \tag{3.19}$$

We note that the local active information storage is defined for every spatiotemporal point (i,n) in the lattice system. We have $A(i,k) = \langle a(i,n,k)\rangle_n$. For systems of homogeneous agents where the PDFs are estimated over all agents, it is appropriate to average over all agents also, giving:

$$A(k) = \langle a(i,n,k)\rangle_{i,n}. \tag{3.20}$$

The average active information storage will always be positive (as for the excess entropy), but is limited by the amount of information that can be used in the next state.

This is, it is bounded above by the average information capacity of a single state (e.g. $\log_2 b$ bits where the agent only takes b discrete states). The *local* active information storage is not bound in this manner however, with larger values indicating that the particular past of an agent provides strong positive information about its next state. Furthermore, the local active information storage can be negative, where the past history of the agent is actually *misinformative* about its next state. Similar to the local excess entropy, an observer is misinformed where the probability of observing the given next state in the context of the past history, $p(x_{n+1} \mid x_n^{(k)})$, is lower than the probability $p(x_{n+1})$ of observing that next state without considering the past.

3.2.2 Active Information Storage and Entropy Rate

The *average* information required to predict the next state for agent X is simply the single cell entropy H_X (Eq. (2.1)). We use the mutual information expansion of Eq. (2.8) to express this entropy in terms of the active information storage and entropy rate estimates[5]:

$$H_{X'} = I_{X';X^{(k)}} + H_{X'|X^{(k)}}, \tag{3.21}$$

$$H_{X'} = A_X(k) + H_{\mu X}(k). \tag{3.22}$$

Logically, we can restate this as: the information to compute or predict a given state is the amount predictable from its past (the active memory) plus the remaining uncertainty after examining this memory.

This equation makes explicit our interpretation of information storage in distributed computation. It is the information in the past that is *observed* to contribute to the computation of the next state. Whether this information is actually causal for that next state, either directly or indirectly through neighbouring agents, is irrelevant for this perspective.[6]

3.2.2.1 Relationship in Local Notation

This relationship can be expressed in local notation also. By way of preliminaries, we represent the *local entropy* as:

$$h_X(n+1) = h(x_{n+1}) = -\log_2 \ p(x_{n+1}), \tag{3.23}$$

[5] These equations are correct not only in the limit $k \to \infty$ but for estimates with any value of $k \geq 1$.

[6] The distinction between causal effect and the perspective of computation is explored in Chap. 4.

and in local lattice notation as:

$$h(i, n+1) = h(x_{i,n+1}) = -\log_2 p(x_{i,n+1}). \tag{3.24}$$

We have $H_X = \langle h_X(n) \rangle_n$, $H(i) = \langle h(i,n) \rangle_n$ and in homogeneous systems $H = \langle h(i,n) \rangle_{i,n}$.

Also, the *local temporal entropy rate* estimate is represented as:

$$h_{\mu X}(n+1, k) = h(x_{n+1} \mid x_n^{(k)}) = -\log_2 p(x_{n+1} \mid x_n^{(k)}). \tag{3.25}$$

The estimate is only completely accurate in the following limit:

$$h_{\mu X}(n+1) = \lim_{k \to \infty} -\log_2 p(x_{n+1} \mid x_n^{(k)}). \tag{3.26}$$

In local lattice notation we can express the estimates as:

$$h_\mu(i, n+1, k) = h(x_{i,n+1} \mid x_{i,n}^{(k)}) = -\log_2 p(x_{i,n+1} \mid x_{i,n}^{(k)}). \tag{3.27}$$

Again, we have $H_{\mu X}(k) = \langle h_{\mu X}(n,k) \rangle_n$, $H_\mu(i,k) = \langle h_\mu(i,n,k) \rangle_n$ and in homogeneous systems $H_\mu(k) = \langle h_\mu(i,n,k) \rangle_{i,n}$. Importantly, since $p(x_{n+1} \mid x_n^{(k)}) \leq 1$ the local temporal entropy rate $h_{\mu X}(n+1,k) \geq 0$.

Then, the relationship between the entropy, active information storage and entropy rate can be expressed in local notation:

$$h_X(n+1) = a_X(n+1,k) + h_{\mu X}(n+1,k), \tag{3.28}$$

$$h(i,n+1) = a(i,n+1,k) + h_\mu(i,n+1,k). \tag{3.29}$$

3.2.2.2 Excess Entropy in Terms of Active Information Storage

It is also possible to express the excess entropy in terms of estimates of the active information storage. We rearrange Eq. (3.22) to get $H_{\mu X}(k) = H_{X'} - A_X(k)$, and then substitute this into Eq. (2.18), getting:

$$E_X = \sum_{k=0}^{\infty} \left[A_X - A_X(k) \right]. \tag{3.30}$$

This expression shows that the excess entropy is the sum of underestimates of the active information storage at each finite history length $k \geq 0$.

This relationship is displayed graphically in Fig. 3.2 in a similar fashion to the plot for the excess entropy in terms of entropy rates in [15].[7] Note that $A(k)$ is non-

[7] Note that our sum and the plot start from $k = 0$, unlike the expressions and plots in [15] which start from $L = 1$. The difference is that we have adopted $k = L - 1$ (as per footnote 4 on p. 4) to

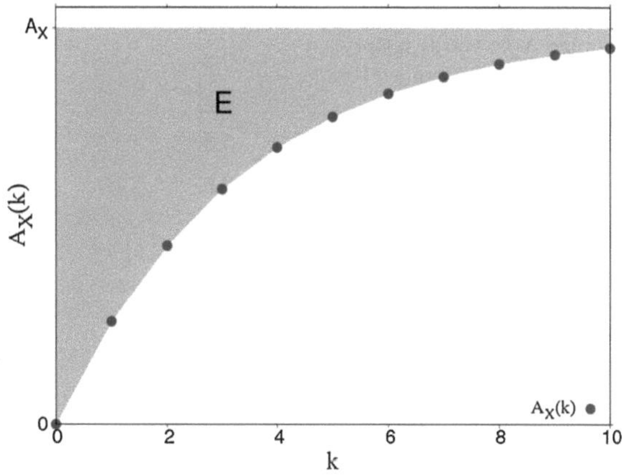

Fig. 3.2 Active information storage convergence: a plot of estimates $A_X(k)$ versus history length k as they converge to the limiting value A_X. The *shaded area* is the excess entropy E. (NB: This figure is reprinted from [7] with permission of Elsevier.)

decreasing with k; this is because increasing the time examined in history widens the scope of temporal correlations that the measure can capture.

3.3 Local Information Storage in Cellular Automata

In this section, we evaluate the local measures within sample runs for the CA rules described in Sect. 2.3.4. To do so, we estimate the required probability distribution functions from CA runs of 10,000 cells, initialised from random states, with 600 time steps retained (after the first 30 time steps were eliminated to allow the CA to settle). Alternatively, for ϕ_{par} we used 1,000 cells with 1,000 time steps retained. Periodic boundary conditions were used. Observations taken at every spatiotemporal point in the CA were used in estimating the required probability distribution functions, since the cells in the CA are homogeneous agents. All conclusions were confirmed by multiple runs from different initial states, and all CA plots were generated using modifications to [16].

In this section we discuss the key results here:

- that one should use as large a history length k as possible to adequately measure information storage (Sects. 3.3.1 and 3.3.2);
- the dominant storage entities in CAs are blinkers and domains (Sect. 3.3.2);
- negative information storage at particles represents the misinformation conveyed by the storage there (Sect. 3.3.3);
- local entropy rate highlights the location of moving particles (Sect. 3.3.7); and
- we discuss minor information storage phenomena (Sects. 3.3.4—3.3.6).

(Footnote 7 continued)
keep a focus on the number of steps k in the past history, which is important for our computational view.

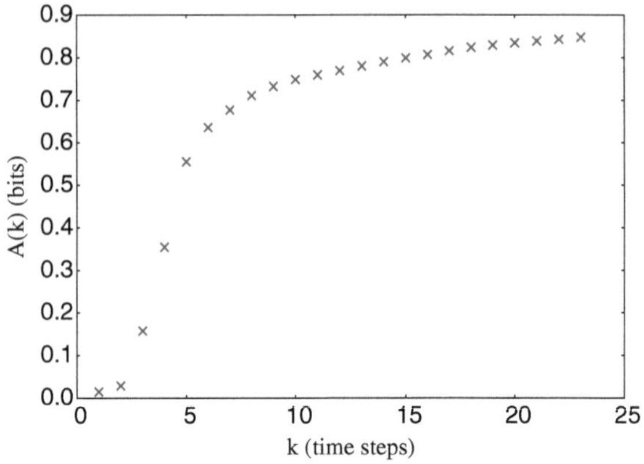

Fig. 3.3 Active information storage $A(k)$ versus history length k for ECA rule 110 (after [7])

3.3.1 Appropriate History Lengths

As previously stated, these measures are only completely correct in the limit $k \to \infty$, however this limit is not computationally achievable. A logical question is what history length k is reasonable to use, noting that setting $k = 1$ is something of a default approach (e.g. for the excess entropy in [17]).

Figure 3.3 presents the average active information storage $A(k)$ in CA rule 110 as a function of history length k. In particular, we observe that using too small a value for k (e.g. $k < 5$ in Fig. 3.3) can lead one to substantially underestimate the information storage. Even in a system as apparently simple as a CA, the default $k = 1$ is clearly inadequate. Obviously in measuring information storage one wants to capture all of the temporal correlations and so use $k \to \infty$. However, the selection of k is limited not only by the amount of history available to examine, but also by the number of observations available for probability distribution function estimation. If k is made too large, the mutual information will be artificially inflated due to under-sampling.

One simple recommended heuristic is to select k to have at least three times as many samples as possible state configurations [18], which would suggest keeping $k \leq 19$ here. More formally however, we select k to have at least M samples on average for each observation in the *typical set* of state configurations [19, 20]. The typical set refers to the set of state configurations where the "sample entropy is close to the true entropy" of that joint state [19], and can be thought of as the set of state configurations likely to be encountered frequently enough to contribute to that entropy. For k length blocks of binary variables, the size of the typical set can be approximated as $2^{h_\mu k}$. For our purposes with $A(k)$, we are considering k length blocks plus the next state, so calculate the typical set for our state configurations as $2^{h_\mu (k+1)}$. With $h_\mu = 0.18$ estimated using $k = 16$ for rule 110 (see Table G.1 in Appendix G), we find the size of the typical set grows much more slowly with k than the set of possible state configurations. This means that fulfilling a desire for $M > 10$ for reasonable accuracy is easily fulfilled for rule 110 and most other rules using $k \leq 18$. Indeed, it is only for rules with $h_\mu \to 1$ (e.g. rule 30, see Table G.1 in Appendix G) that M even approaches 10 with $k \leq 18$; for most rules $k \leq 18$ results in a much larger average number of samples $M >> 10$ and therefore larger accuracy. We elect to continue our CA investigations with the more strict condition $k \leq 16$ however, since this satisfies the $M > 10$ condition for all ECAs for the measures we consider in later sections (since these measures consider a higher-dimensional joint space of $k + 3$ variables).

Figure 3.3 suggests that the majority of the information storage for rule 110 is captured with $k \geq 7$ or so, however this examination of the average values of $A(k)$ does not show explicitly why this is the case. Furthermore, these average values tell us nothing about whether blinkers are dominant information storage structures, and if so whether the information storage in them has been captured at these history lengths. To understand these issues, we begin to examine the information storage on a local scale in the next section.

3.3.2 *Periodic Blinker and Domain Processes as Dominant Storage*

We begin by examining the local profiles for rules 54 and 110, which are known to contain regular gliders against periodic background domains. For the CA evolutions in Figs. 3.4a and 3.5a, the local profiles of $e(i, n, k = 8)$ are displayed in Figs. 3.4b and 3.5b, and the local profiles of $a(i, n, k = 16)$ in Figs. 3.4c and 3.5c.[8]

It is quite clear that positive information storage is concentrated in the vertical gliders or blinkers, and the domain regions. As expected, these results provide quantitative evidence that the **blinkers are the dominant information storage entities**. This is because the cell states in the blinkers are strongly predictable from their past history, since they are temporally periodic. It is only the local profiles that demonstrate the strong information storage at these entities though. That **the domain regions for these rules also contain significant information storage** should not be surprising, since these too are periodic and so their past does indeed store information about their future. In fact, the local values for each measure form spatially and temporally periodic patterns in these regions, due to the underlying periodicities exhibited there.

Yet **the local active information storage and local excess entropy yield subtly different results here**. While $a(i, n, k = 16)$ indicates a similar amount of stored information in use to compute each space-time point in both the domain and blinker areas, $e(i, n, k = 8)$ reveals a larger *total* amount of information is stored in the blinkers. For the blinkers known as α and β in rule 54 [22] this is because the temporal sequences of the centre columns of the blinkers (0–0–0–1, with $e(i, n, k = 8)$ in the range 5.01–5.32 bits) are more complex than those in the domain (0–0–1–1 and 0–1, with $e(i, n, k = 8)$ in the range 1.94–3.22 bits), even where they are of the same period. In principle, we could define a threshold i_B to differentiate between the

---

[8] Sub-figures (e)–(i) of these figures will be discussed in later chapters.

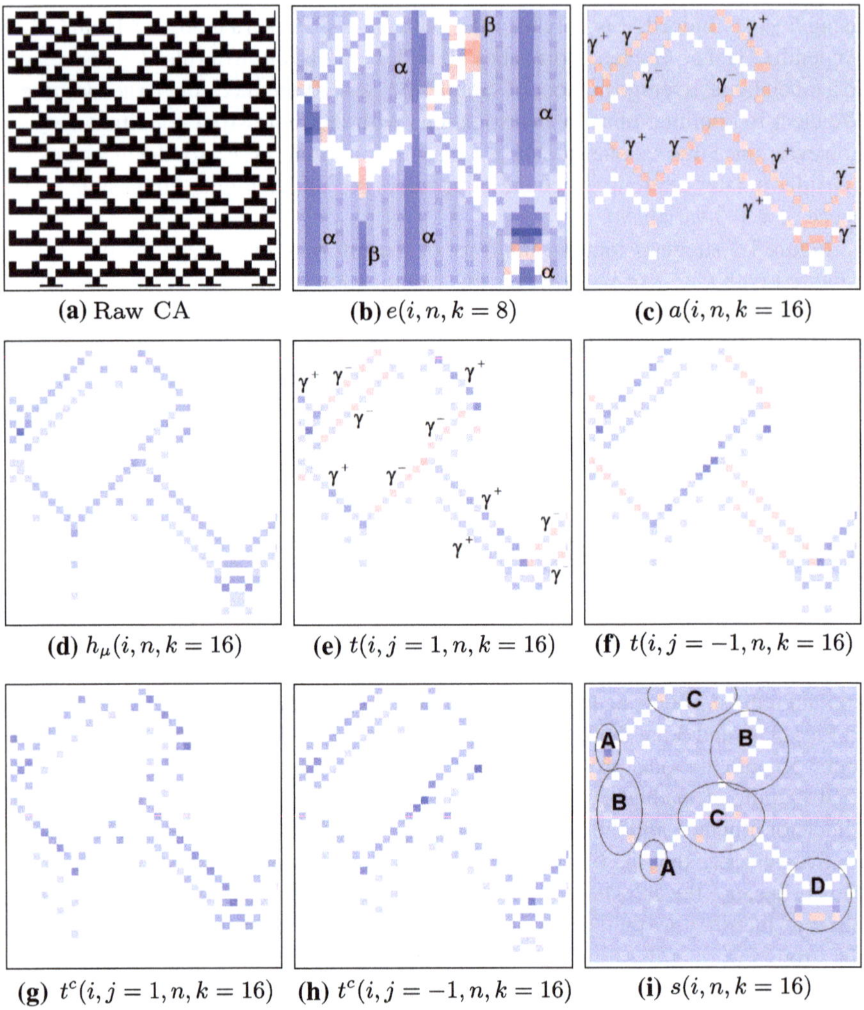

Fig. 3.4 **a** Local information dynamics in **rule 54** (35 time steps displayed for 35 cells, time increases down the page for all CA plots). **All profiles are discretised into 16 levels, with *blue* for positive values and *red* for negative.** **b** Local excess entropy, max. 11.79 bits, min. −12.35 bits; **c** Local active information, max. 1.07 bits, min. −12.27 bits; **d** Local temporal entropy rate, max. 13.20 bits, min. 0.00 bits; Local apparent transfer entropy: **e** one cell to the right, max. 7.93 bits, min. −4.04 bits, **f** one cell to the *left*, max. 7.93 bits, min. −4.21 bits; Local complete transfer entropy: **g** one cell to the *right*, max. 9.22 bits, min. 0.00 bits, **h** one cell to the *left*, max. 9.48 bits, min. 0.00 bits; **i** Local separable information, max. 8.40 bits, min. −5.27 bits. Note that the quantities in (**e**)–(**i**) will be described in later chapters. (NB: Fig. 3.4a, c, e, f and i Reprinted with permission from Lizier et al. [21] Copyright 2010, American Institute of Physics.)

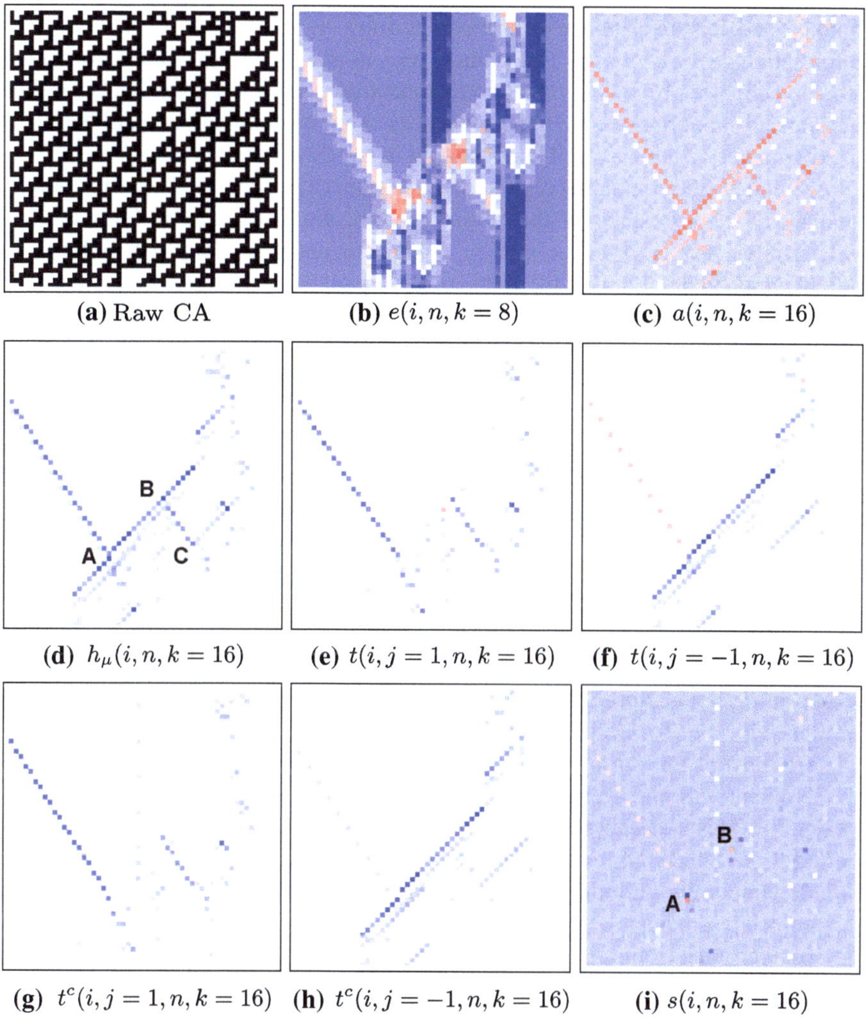

Fig. 3.5 **a** Local information dynamics in **rule 110** (55 time steps displayed for 55 cells). Cells are coloured *blue* for positive values and *red* for negative. **b** Local excess entropy, max. 10.01 bits, min. −10.35 bits; **c** Local active information, max. 1.22 bits, min. −9.21 bits; **d** Local temporal entropy rate, max. 10.43 bits, min. 0.00 bits; Local apparent transfer entropy: **e** one cell to the *right*, max. 9.99 bits, min. −5.56 bits, **f** one cell to the *left*, max. 10.43 bits, min. −6.01 bits; Local complete transfer entropy: **g** one cell to the *right*, max. 9.99 bits, min. 0.00 bits, **h** one cell to the *left*, max. 10.05 bits, min. 0.00 bits; **i** Local separable information, max. 5.47 bits, min. −5.20 bits (*black*). (NB: Fig. 3.5a, c, e, f and i Reprinted with permission from Lizier et al. [21] Copyright 2010, American Institute of Physics.)

blinkers and domain using the local excess entropy. Note that we have the total stored information $e(i, n, k = 8) > 1$ bit in these regions due to the distributed information storage supported by bidirectional communication (as discussed in Sect. 3.1). This mechanism supports these periodic sequences being longer than two time steps (the maximum period a binary cell could sustain in isolation).

As another rule containing regular gliders against a periodic background domain, analysis of the raw states of the density classification rule ϕ_{par} in Fig. 3.6a provides similar results for $e(i, n, k = 5)$ in Fig. 3.6b and $a(i, n, k = 10)$ in Fig. 3.6c here. One distinction is that the blinkers here contain no more stored information than the domains, since they are no more complicated. Another distinction is that the profiles of active information storage and excess entropy are more similar here than for the other rules: this is because the domain patterns here are very simple (with a period of at most 2), so $a(i, n, k = 10)$ detects almost all of the total information storage that is used in the future (the excess entropy) here even though it only examines a single future step. Importantly, we confirm the information storage capability of the blinkers and domains in this human understandable computation.

To further investigate the appropriate history length k for use with the information storage measures, we examine the profiles of $a(i, n, k = 1)$ and $a(i, n, k = 7)$ for rule 110 in Fig. 3.7. As per the low average value for $A(k = 1)$ in Figs. 3.3, 3.7b demonstrates that the use of $k = 1$ is inadequate here since it does not capture the strong information storage in the gliders and domain regions that we see for the profiles with $k = 16$. On the other hand, Fig. 3.7c shows that the use of $k = 7$ does capture most of this strong information storage (compare to $k = 16$ in Fig. 3.5c), in alignment with the average value for $A(k = 7)$ approaching the limiting value in Fig. 3.3. This is because the blinker and domain regions for rule 110 are both of period 7. To understand why setting k at this period is effective, consider first an *infinite* temporally periodic process with period p. The next state of that process is completely predictable from its (infinite) past. In fact, the number of past states an observer must examine to correctly determine the next state is limited by p (as per the synchronisation time in [23]). Using $k > p - 1$ does not add any extra information about the next state than is already contained in the $p - 1$ previous states. However, using $k < p - 1$ may not provide sufficient information for the prediction. Using $k = p - 1$ is a sufficient (Markovian) condition for infinitely periodic processes. **Where a process contains *punctuated* periodic sequences** (e.g. the periodic blinkers and domains in a single cell's time series here), **setting $k = p-1$ will capture the information storage related to the period of these sequences and is a useful minimum value**. However, it will still ignore important longer-range correlations (e.g. encountering one type of glider in the near past may be strongly predictive of encountering a different type of glider in the near future). There is no general limit on the range of such self-influence, so in theory the limit $k \to \infty$ should be taken in measuring these quantities. This is why $k = 7$ captures much, but not all of the active information storage for rule 110.

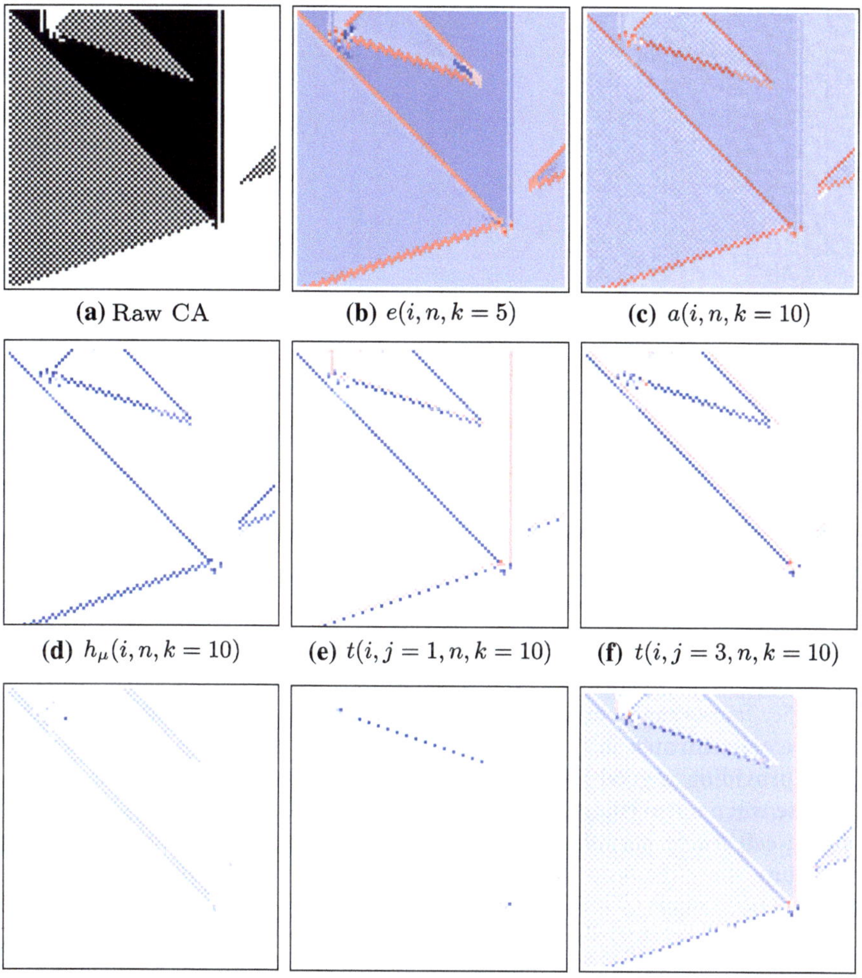

(a) Raw CA (b) $e(i, n, k = 5)$ (c) $a(i, n, k = 10)$

(d) $h_{\mu}(i, n, k = 10)$ (e) $t(i, j = 1, n, k = 10)$ (f) $t(i, j = 3, n, k = 10)$

(g) $t^c(i, j = 1, n, k = 10)$ (h) $t^c(i, j = 3, n, k = 10)$ (i) $s(i, n, k = 10)$

Fig. 3.6 **a** Local information dynamics in $r = 3$ **rule** $\phi_{\mathbf{par}}$ (86 time steps displayed for 86 cells). Cells are coloured *blue* for positive values and *red* for negative. **b** Local excess entropy, max. 11.76 bits, min. −10.35 bits; **c** Local active information, max. 1.52 bits, min. −9.41 bits; **d** Local temporal entropy rate, max. 10.92 bits, min. 0.00 bits; Local apparent transfer entropy: **e** one cell to the *right*, max. 10.45 bits, min. −8.06 bits, and **f** three cells to the *right*, positive values only, max. 9.24 bits, min. −8.60 bits; Local complete transfer entropy: **g** one cell to the *right*, max. 10.43 bits, min. 0.00 bits, and **h** three cells to the *left*, max. 10.86 bits, min. 0.00 bits; **i** Local separable information, max. 29.26 bits, min. −18.68 bits

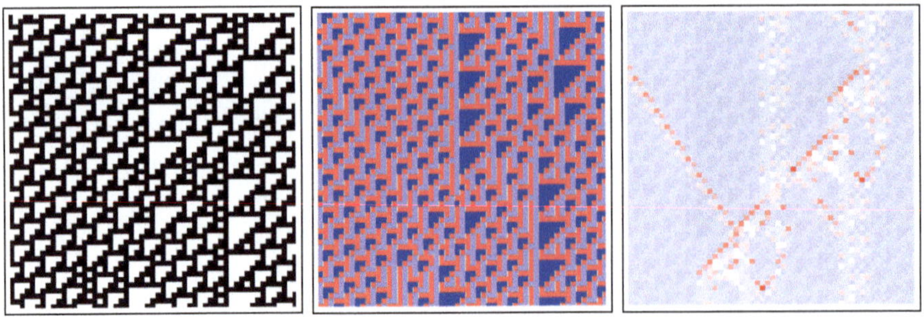

(a) Raw CA (b) $a(i, n, k = 1)$ (c) $a(i, n, k = 7)$

Fig. 3.7 Local active information storage with short history lengths in rule 110. **a** 35 time steps displayed for 35 cells, time increases down the page for all CA plots. Cells are coloured *blue* for positive values and *red* for negative. **b** With $k = 1$, max. 0.24 bits, min. −0.21 bits; **c** with $k = 7$, max. 1.22 bits, min. −5.04 bits. (NB: Fig. 3.7a Reprinted with permission from Lizier et al. [21] Copyright 2010, American Institute of Physics.)

3.3.3 Negative Informative Storage as Misinformation at Particles

Negative values of $a(i, n, k = 16)$ for rules 54 and 110 are also displayed in Figs. 3.4c and 3.5c. Interestingly, **negative local components of active information storage are concentrated in the travelling glider areas** (e.g. γ^+ and γ^- for rule 54 [22]), **providing a good spatiotemporal filter of the glider structure. This is because when a travelling glider is encountered at a given cell, the past history of that cell** (being part of the background domain) **is *misinformative* about the next state**, since the domain sequence was more likely to continue than be interrupted. For example, see the marked positions of the γ gliders in Fig. 3.8. There we have $p(x_{n+1} \mid x_n^{(k=16)}) = 0.25$ and $p(x_{n+1}) = 0.52$: since the next state occurs relatively infrequently after the given history, that history provides a misinformative $a(n, k = 16) = -1.09$ bits about the next state. This is juxtaposed with the points four time steps before those marked "x", which have the same history $x_n^{(k=16)}$ but remain part of the domain. There we have $p(x_{n+1} \mid x_n^{(k=16)}) = 0.75$ and $p(x_{n+1}) = 0.48$ giving $a(n, k = 16) = 0.66$ bits, quantifying the positive information storage there.

Note that the points with misinformative storage are not necessarily those selected by other filtering techniques as part of the gliders. For example, the finite state transducers technique from computational mechanics (using left to right spatial scanning by convention) [24] would identify points 3 cells to the right of those marked "x" as part of the γ^+ glider. While that technique has the perspective of spatial pattern recognition, we take the temporal perspective of unfolding computation.

Fig. 3.8 Close up of raw states of rule 54. "x" and "+" mark some positions in the γ^+ and γ^- gliders respectively. Note their point of coincidence in collision type "A", with "$*$" marking what initially appears to be the collision point and "o" marking the subsequent non-trivial information modification as detected using $s(i, n, k = 16) < 0$ (see Chap. 5). (Reprinted with permission from Lizier et al. [21] Copyright 2010, American Institute of Physics.)

The local excess entropy also produced some negative values around travelling gliders (see Figs. 3.4b and 3.5b), though these were far less localised on the gliders themselves and less consistent in occurrence than for the local active information storage. This is because the local excess entropy, as measure of *total* information storage into the future, is more loosely tied to the dynamics at the given spatiotemporal point. The effect of a glider encounter on $e(i, n, k)$ is smeared out in time, and in fact the dynamics may store more positive information in total than the misinformation encountered at the specific location of the glider. For example, parallel glider pairs in Fig. 3.4b have positive total information storage, since a glider encounter becomes much more likely in the wake of a previous glider.

3.3.4 Particles Create New Information Storage

There is also strong positive information storage in the "wake" of the more complex gliders in rule 110 (e.g. see the glider at top left of Fig. 3.5b, c). This indicates that while the leading edge of the gliders cause the cell states to become unpredictable from their past, the subsequent activity (before a domain pattern is established) is predictable given the glider encounter. **The leading edge of the gliders can thus be seen to store information in the cell about its new behaviour**. The presence of this information storage is shown by both measures, although the relative strength of the total information storage is again revealed only by the local excess entropy. We will observe a similar creation of new information storage by domain walls in rule 18 in Sect. 3.3.6.

3.3.5 Structured Information Storage in Domain of Rule 18

There is also interesting information storage structure in ECA rule 18, which contains domain walls against a seemingly irregular background domain. The local profiles for $e(i, n, k = 8)$ and $a(i, n, k = 16)$ are plotted in Fig. 3.9b, c for the raw states of rule 18 displayed in Fig. 3.9a. In contrast to rules 54 and 110, the background domain for rule 18 contains points with both positive and negative local active information storage. Considering these components together, we observe a pattern to the background domain of spatial and temporal period 2 corresponding to the period-2 ϵ-machine generated to recognise the background domain for ECA rule 18 by Hanson and Crutchfield [25]. Every second site in the domain is a “0”, and contains a small positive $a(i, n, k = 16)$ ($\approx$0.43 to 0.47 bits); information storage of this primary temporal phase of the period is sufficient to predict the next state here. The alternate site is either a “0” or a “1”, and contains either a small negative $a(i, n, k = 16)$ at the “0” sites (≈ -0.45 to -0.61 bits) or a larger positive $a(i, n, k = 16)$ at the “1” sites ($\approx$0.98 to 1.09 bits). Information storage of the cell being in the alternate temporal phase is strongly in use or active in computing the “1” sites, since the “1” sites only occur in the alternate phase. However, the information storage indicating the alternate temporal phase is misleading in computing the “0” sites since they occur more frequently with the primary phase. Indeed, encountering a “0” at the alternate sites creates ambiguity in the future (since it makes determination of the phase more difficult) so in this sense it can be seen as detracting from the overall storage. The background domain should contain a consistent level of excess entropy at 1 bit to store the temporal phase information, and this occurs for most points.[9] Again, this resembles a smearing out of the local periodicity of the storage in use, and **highlights the subtle differences between the excess entropy and active information storage**.

3.3.6 Misinformation and New Storage Creation by Domain Walls

The domain walls in rule 18 are points where the spatiotemporal domain pattern is violated. **Strong negative components of the local active information storage reveal the temporal violations, which occur when the domain wall moves** or travels into a new cell and the past of that cell cannot then predict the next state successfully. This misinformation is analogous to our observations for regular gliders in rules 54 and 110 in Sect. 3.3.3. Importantly, these negative values of $a(i, n, k = 16)$ (which are less than -2.5 bits) are much stronger than those in the background domain, and are strongly localised on the domain walls. Again, the negative components of $a(i, n, k = 16)$ appear to be a useful filter for moving coherent spatiotemporal structure.

---

9 The exceptions are where long temporal chains of 0’s occur, disturbing the memory of the phase due to finite-k effects.

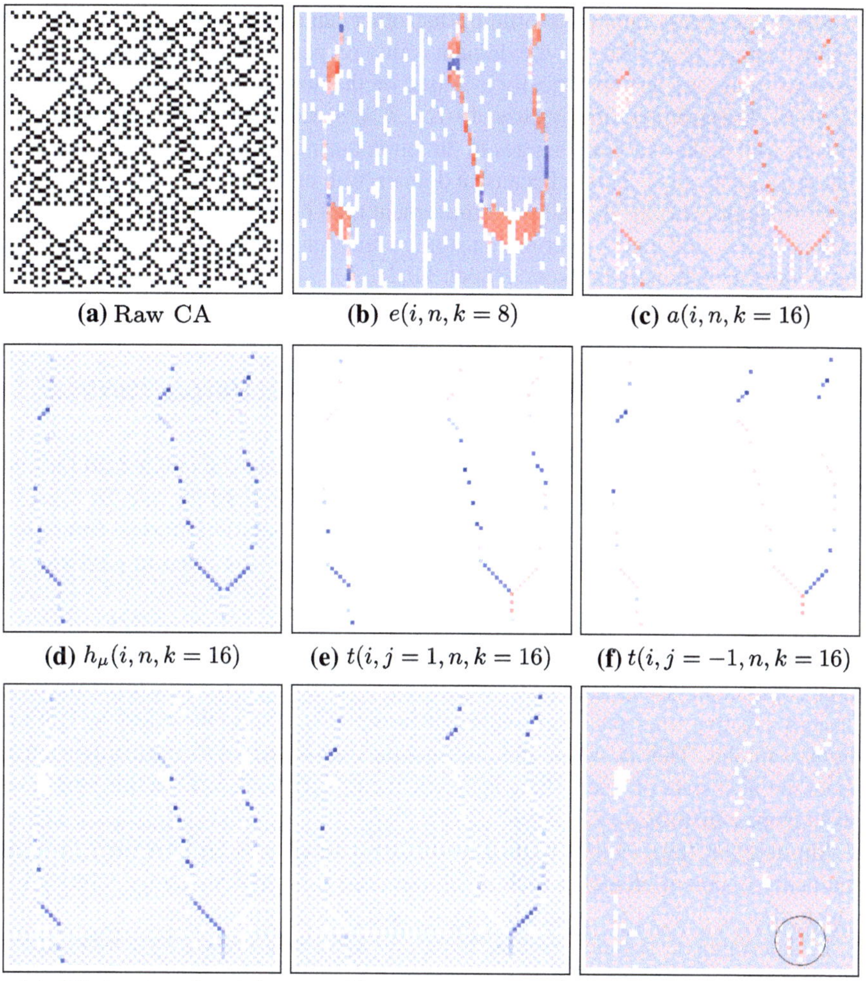

(a) Raw CA (b) $e(i, n, k = 8)$ (c) $a(i, n, k = 16)$

(d) $h_\mu(i, n, k = 16)$ (e) $t(i, j = 1, n, k = 16)$ (f) $t(i, j = -1, n, k = 16)$

(g) $t^c(i, j = 1, n, k = 16)$ (h) $t^c(i, j = -1, n, k = 16)$ (i) $s(i, n, k = 16)$

Fig. 3.9 **a** Local information dynamics in **rule 18** (67 time steps displayed for 67 cells). Cells are coloured *blue* for positive values and *red* for negative. **b** Local excess entropy, max. 4.62 bits, min. −8.65 bits; **c** Local active information, max. 1.98 bits, min. −9.92 bits; **d** Local temporal entropy rate, max. 11.90 bits, min. 0.00 bits; Local apparent transfer entropy: **e** one cell to the *right*, max. 11.90 bits, min. −7.44 bits, **f** one cell to the *left*, max. 11.90 bits, min. −7.30 bits; Local complete transfer entropy: **g** one cell to the *right*, max. 13.48 bits, min. 0.00 bits, **h** one cell to the *left*, max. 12.48 bits, min. 0.00 bits; **i** Local separable information, max. 1.98 bits, min. −14.37 bits. (NB: Fig. 3.9a, c, e, f and i Reprinted with permission from Lizier et al. [21] Copyright 2010, American Institute of Physics.)

The local excess entropy profile on the other hand contains both positive and negative values (Fig. 3.9b) for the domain walls. As per the results for gliders, these negative values are less specifically localised on the domain walls than observed for $a(i, n, k)$. The strong positive values of $e(i, n, k = 8)$ are observed to occur where the domain wall makes several changes of direction during the k steps but is somewhat stationary on average. This is because a domain wall encounter is much more likely in the wake of previous domain wall movement than elsewhere in the CA. This has analogies to both the new information storage creation by gliders in Sect. 3.3.3, and the storage in stationary blinkers in Sect. 3.3.2.

3.3.7 Local Temporal Entropy Rate Highlights Moving Particles

The local temporal entropy rate profiles $h_\mu(i, n, k)$ are displayed in Fig. 3.4d for rule 54, Fig. 3.5d for rule 110, Fig. 3.6d for rule ϕ_{par} and Fig. 3.9d for rule 18. Clearly **these local temporal entropy rate profiles are useful spatiotemporal filters for moving emergent structure, since they highlight all of the moving particles in each system**.

In fact, these profiles are quite similar to those of the negative values of local active information storage.[10] This is not surprising since $h_\mu(i, n, k)$ and $a(i, n, k)$ are seen to be complementary in Eq. (3.28). Where $a(i, n, k)$ is negative, $h_\mu(i, n, k)$ must be strongly positive since the local single cell entropy $h(i, n)$ averages close to 1 bit for these examples. That is, where the information storage $a(i, n, k)$ is misinformative about the next state of a cell, there is a high uncertainty $h_\mu(i, n, k)$ in this next state given its past history.

Similarly, $h_\mu(i, n, k)$ only highlights moving particles: as we have already seen, stationary coherent elements such as blinkers are information storage entities for which there is little to no uncertainty in the next state given the past.

In deterministic systems such as CAs, any extra information $h_\mu(i, n, k)$ about the next state *must* come from the neighbouring cells. **The temporal entropy rate therefore represents a *collective information* transfer from the neighbours in deterministic systems**. Importantly though, it cannot identify individual sources of information transfer: this will be investigated in Chap. 4.

3.3.8 Absence of Coherent Information Storage Structure

Finally, we examine ECA rule 22, suggested to have infinite collective excess entropy [11, 12] but without any known coherent structural elements [26]. For the raw states of rule 22 displayed in Fig. 3.10a, the profiles of $e(i, n, k = 8)$ and $a(i, n, k = 16)$ are shown in Fig. 3.10b, c. While storage is certainly observed to occur, **these plots**

[10] Recall though from Sect. 3.2.2 that $h_\mu(i, n, k)$ itself is never negative.

(a) Raw CA (b) $e(i, n, k = 8)$ (c) $a(i, n, k = 16)$

(d) $h_\mu(i, n, k = 16)$ (e) $t(i, j = 1, n, k = 16)$ (f) $t(i, j = -1, n, k = 16)$

(g) $t^c(i, j = 1, n, k = 16)$ (h) $t^c(i, j = -1, n, k = 16)$ (i) $s(i, n, k = 16)$

Fig. 3.10 **a** Local information dynamics in **rule 22** (67 time steps displayed for 67 cells). Cells are coloured *blue* for positive values and *red* for negative. **b** Local excess entropy, max. 4.49 bits, min. −8.76 bits; **c** Local active information: max. 1.51 bits, min. −8.17 bits; **d** Local temporal entropy rate, max. 9.68 bits, min. 0.00 bits; Local apparent transfer entropy: **e** one cell to the *right*, max. 9.68 bits, min. −7.05 bits, **f** one cell to the *left*, max. 9.68 bits, min. −7.10 bits; Local complete transfer entropy: **g** one cell to the *right*, max. 10.68 bits, min. 0.00 bits, **h** one cell to the *left*, max. 10.68 bits, min. 0.00 bits; **i** Local separable information, max. 5.03 bits, min. −14.44 bits. (NB: Fig. 3.10a, c and f are reprinted from [27] with permission of Springer.)

provide evidence that there is no coherent structure to the information storage for rule 22. This is another clear example of the utility of examining local information dynamics over ensemble values.

We note that the chaotic rule 30 produces profiles with a similar lack of coherent structure (see Fig. 3.11b, c).

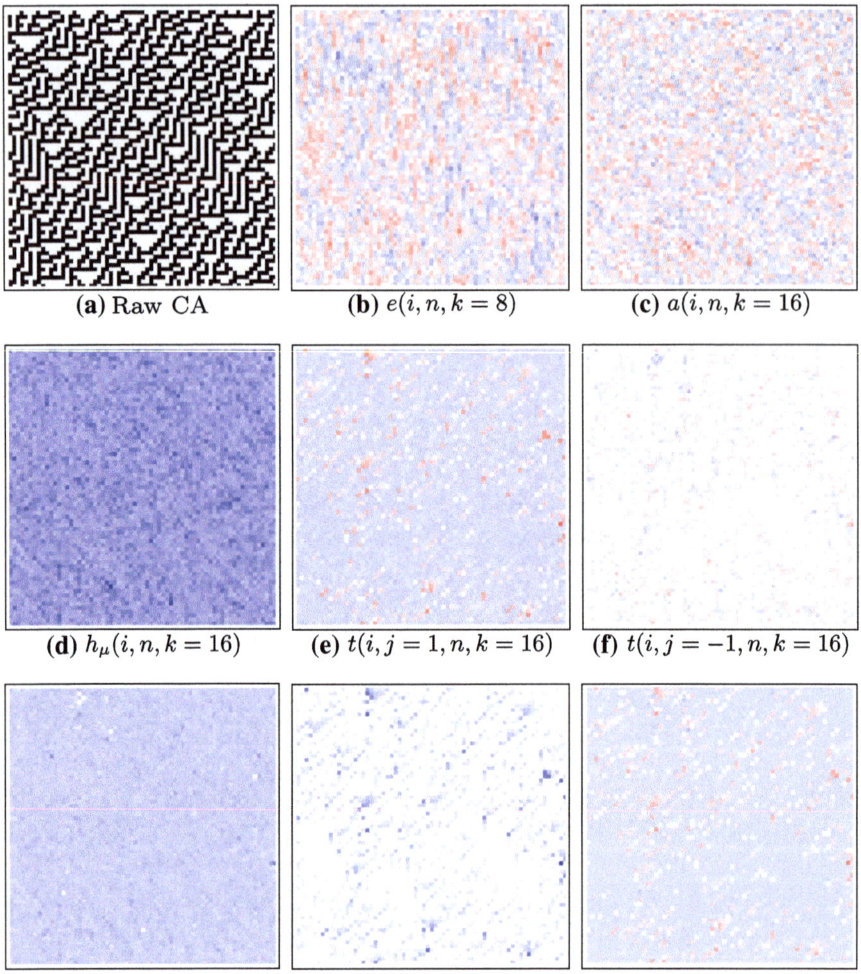

(a) Raw CA (b) $e(i, n, k = 8)$ (c) $a(i, n, k = 16)$

(d) $h_\mu(i, n, k = 16)$ (e) $t(i, j = 1, n, k = 16)$ (f) $t(i, j = -1, n, k = 16)$

(g) $t^c(i, j = 1, n, k = 16)$ (h) $t^c(i, j = -1, n, k = 16)$ (i) $s(i, n, k = 16)$

Fig. 3.11 **a** Local information dynamics in **rule 30** (67 time steps displayed for 67 cells). Cells are coloured *blue* for positive values and *red* for negative. **b** Local excess entropy, max. 0.59 bits, min. −0.79 bits; **c** Local active information: max. 0.54 bits, min. −0.87 bits; **d** Local temporal entropy rate, max. 1.87 bits, min. 0.00 bits; Local apparent transfer entropy: **e** one cell to the *right*, max. 1.87 bits, min. −4.95 bits, **f** one cell to the *left*, max. 1.81 bits, min. −2.74 bits; Local complete transfer entropy: **g** one cell to the *right*, max. 3.81 bits, min. 0.00 bits, **h** one cell to the *left*, max. 6.13 bits, min. 0.00 bits; **i** Local separable information, max. 2.81 bits, min. −6.90 bits. (NB: Fig. 3.11a, c and f are reprinted from [27] with permission of Springer.)

Table 3.1 Local measures relevant to information storage

Measure	Local definition	Equation number
Excess entropy	$e_X(n+1,k) = i(x_n^{(k)}; x_{n+1}^{(k^+)})$	Eq. (3.3)
Active information storage	$a_X(n+1,k) = i(x_n^{(k)}; x_{n+1})$	Eq. (3.17)
Entropy rate	$h_{\mu X}(n+1,k) = h(x_{n+1} \| x_n^{(k)})$	Eq. (3.25)

Table 3.2 Examples of emergent structures in cellular automata with specific information storage properties

$a_X(n) > 0$	$e_X(n) > i_B$	$e_X(n) < i_B$
	Blinkers, stationary domain walls (Sects. 3.3.2 and 3.3.6)	Periodic domain (Sect. 3.3.2)
$a_X(n) < 0$	Gliders, moving domain walls (Sects. 3.3.3 and 3.3.6)	

3.4 Summary

In this chapter we have introduced and contrasted the local excess entropy and local active information storage in Sects. 3.1 and 3.2. Their definitions are displayed in Table 3.1. We have demonstrated that they provide complementary insights into information storage dynamics, because the excess entropy measures total storage while the active information storage measures storage in use in computing the next state. As such, their results are often similar in general, but do reveal subtly different aspects of the dynamics.

Importantly, in Sect. 3.3 we showed that both provide the first quantitative evidence that blinkers and domains are dominant information storage entities in cellular automata. In particular, the excess entropy revealed that the blinkers in the CAs investigated here stored more information in total. (See a summary of the application to CAs in Table 3.2). While both measures provide useful insights, the local active information storage is the most useful in a *real-time* sense, since calculation of the local excess entropy requires knowledge of the dynamics an arbitrary distance into the future.[11] Also, it also provides the most specifically *localised* insights, including highlighting moving elements of coherent spatiotemporal structure.

Furthermore as hinted by its relationship with the temporal entropy rate, the focus of the active information storage on computation of the *next state* of a process is particularly important in understanding how stored information interacts with information transfer in information processing. This being said, it is not capable of identifying

[11] As described in footnote 4 on p. 4, while there are alternative formulations of the local excess entropy which can be computed from past observations alone, they cannot be interpreted as the total information storage at the given time point. A similar concept would be the partial localisation (see Appendix A) $I(x_n^{(k)}; X^{(k^+)})$, which quantifies how much information from the past is *likely* to be used in the future.

the information *source* of moving coherent structures; for this, we turn our attention to measuring information transfer in the next chapter.

References

1. M.G. Kitzbichler, M.L. Smith, S.R. Christensen, E. Bullmore, Broadband criticality of human brain network synchronization. PLoS Comput. Biol. **5**(3), e1000314 (2009)
2. J. Boedecker, O. Obst, N.M. Mayer, M. Asada, Initialization and self-organized optimization of recurrent neural network connectivity. HFSP J. **3**(5), 340–349 (2009)
3. R. Morgado, M. Ciesla, L. Longa, F.A. Oliveira, Synchronization in the presence of memory. Europhys. Lett. **79**(1), 10002 (2007)
4. M. Prokopenko, V. Gerasimov, I. Tanev, Evolving spatiotemporal coordination in a modular robotic system, in *Proceedings of the Ninth International Conference on the Simulation of Adaptive Behavior (SAB'06)*, Rome, ed. by S. Nolfi, G. Baldassarre, R. Calabretta, J. Hallam, D. Marocco, J.-A. Meyer, D. Parisi. Lecture Notes in Artificial Intelligence, vol. 4095 (Springer, Berlin, 2006), pp. 548–559
5. K.I. Goh, A.L. Barabási, Burstiness and memory in complex systems. Europhys. Lett. **81**(4), 48002 (2008)
6. J.T. Lizier, M. Prokopenko, A.Y. Zomaya, Detecting non-trivial computation in complex dynamics, in *Proceedings of the 9th European Conference on Artificial Life* (*ECAL, 2007*) Lisbon, Portugal, ed. by F. Almeida e Costa, L.M. Rocha, E. Costa, I. Harvey, A. Coutinho. Lecture Notes in Artificial Intelligence, vol. 4648 (Springer, Berlin, 2007), pp. 895–904
7. J.T. Lizier, M. Prokopenko, A.Y. Zomaya, Local measures of information storage in complex distributed computation. Inf. Sci. **208**, 39–54 (2012)
8. C.G. Langton, Computation at the edge of chaos: phase transitions and emergent computation. Physica D **42**(1–3), 12–37 (1990)
9. A.S. Klyubin, D. Polani, C.L. Nehaniv, Tracking information flow through the environment: simple cases of stigmergy, in *Proceedings of the Ninth International Conference on the Simulation and Synthesis of Living Systems* (*ALife IX*), Boston, ed. by J. Pollack, M. Bedau, P. Husbands, T. Ikegami, R.A. Watson (MIT Press, Cambridge, 2004), pp. 563–568
10. I. Couzin, R. James, D. Croft, J. Krause, Social organization and information transfer in schooling fishes, ed. by B.C.K. Laland, J. Krause, in *Fish Cognition and Behavior*, ser. Fish and Aquatic Resources (Blackwell Publishing, Boston, 2006), pp. 166–185
11. P. Grassberger, Toward a quantitative theory of self-generated complexity. Int. J. Theor. Phys. **25**(9), 907–938 (1986)
12. P. Grassberger, Long-range effects in an elementary cellular automaton. J. Stat. Phys. **45**(1–2), 27–39 (1986)
13. K. Lindgren, M.G. Nordahl, Complexity measures and cellular automata. Complex Syst. **2**(4), 409–440 (1988)
14. C.R. Shalizi, Causal architecture, complexity and self-organization in time series and cellular automata. Ph.D. Dissertation, University of Wisconsin-Madison, 2001
15. J.P. Crutchfield, D.P. Feldman, Regularities unseen, randomness observed: levels of entropy convergence. Chaos **13**(1), 25–54 (2003)
16. M. Wójtowicz, Java Cellebration v. 1.50, Online software (2002), http://psoup.math.wisc.edu/mcell/mjcell/mjcell.html
17. N. Ay, N. Bertschinger, R. Der, F. Güttler, E. Olbrich, Predictive information and explorative behavior of autonomous robots. Eur. Phys. J. B **63**(3), 329–339 (2008)
18. M. Lungarella, T. Pegors, D. Bulwinkle, O. Sporns, Methods for quantifying the informational structure of sensory and motor data. Neuroinformatics **3**(3), 243–262 (2005)
19. T.M. Cover, J.A. Thomas, *Elements of Information Theory* (Wiley, New York, 1991)

20. K. Marton, P.C. Shields, Entropy and the consistent estimation of joint distributions. Ann. Probab. **22**(2), 960–977 (1994)
21. J.T. Lizier, M. Prokopenko, A.Y. Zomaya, Information modification and particle collisions in distributed computation. Chaos **20**(3), 037109 (2010)
22. W. Hordijk, C.R. Shalizi, J.P. Crutchfield, Upper bound on the products of particle interactions in cellular automata. Physica D **154**(3–4), 240–258 (2001)
23. D.P. Feldman, J.P. Crutchfield, Synchronizing to periodicity: the transient information and synchronization time of periodic sequences. Adv. Complex Syst. **7**(3–4), 329–355 (2004)
24. J.E. Hanson, J.P. Crutchfield, Computational mechanics of cellular automata: an example. Physica D **103**(1–4), 169–189 (1997)
25. J.E. Hanson, J.P. Crutchfield, The attractor-basin portait of a cellular automaton. J. Stat. Phys. **66**, 1415–1462 (1992)
26. C.R. Shalizi, R. Haslinger, J.-B. Rouquier, K.L. Klinkner, C. Moore, Automatic filters for the detection of coherent structure in spatiotemporal systems. Phys. Rev. E **73**(3), 036104 (2006)
27. J.T. Lizier, M. Prokopenko, A.Y. Zomaya, Coherent information structure in complex computation, Theory Biosci. **131**(3), 193–203 (2012), doi:10.1007/s12064-011-0145-9

Chapter 4
Information Transfer

Information transfer is widely considered to be a vital component of complex non-linear behaviour in spatiotemporal systems, for example in: particles in cellular automata (CAs) [1–7], self-organisation caused by dipole-dipole interactions in microtubules [8], soliton dynamics and collisions [9], wave-fragment propagation in Belousov-Zhabotinsky media [10], waves of turning movement in flocking or schooling behaviour [11], solid-state phase transitions in crystals [12], influence of intelligent agents over their environments [13], and inducing emergent neural structure [14].

As described in Sect. 2.2.4.2, the very nature of information transfer in complex systems is a popular topic itself, for example in the conflicting suggestions that information transfer is maximised in complex dynamics [15, 16], or alternatively at an intermediate level with maximisation leading to chaos [5, 17]. Yet while the literature contains many measures of complexity (e.g. [6, 18]), quantitative studies of information transfer are comparatively rare.

Defining information transfer as the dependence of the next state of the receiver on the previous state of the source [19] is typical, though it is incomplete according to Schreiber's criteria [20] requiring the definition to be both directional and dynamic. Here, *in theory* we accept Schreiber's definition [20] of (predictive) information transfer as the average information contained in the source about the next state of the destination that was not already contained in the destination's past. This definition results in the measure for information transfer known as *transfer entropy* [20], quantifying "the statistical coherence between systems evolving in time" in a directional and dynamic manner.

However, neither this nor other suggested measures of information transfer have been applied to specific channels or at specific spatiotemporal points in a CA,[1] nor quantitatively demonstrated that particles (either gliders or domain walls) are in fact

[1] A rudimentary attempt was made via mutual information in [5], however we show that this is a symmetric measure not capturing directional transfer. We also note the inferences made regarding the spread of information in ECAs [21], however the use of spatial excess entropy (or effective measure complexity) here also remains a symmetric measure of statically shared information.

J. T. Lizier, *The Local Information Dynamics of Distributed Computation in Complex Systems*, Springer Theses, DOI: 10.1007/978-3-642-32952-4_4,

© Springer-Verlag Berlin Heidelberg 2013

information transfer entities. We hypothesise that *application of a measure of local information transfer into each spatiotemporal point in CAs would reveal particles as the dominant information transfer entities*. This would demonstrate that the measure captures popularly-understood, recognised instances of information transfer, and is a key requirement for it to be accepted as measuring this concept *in practice*. We note that these instances are local in space and time and to be investigated require a local measure of information transfer.

Here then, we derive a measure of *local* information transfer from the transfer entropy in Sect. 4.1.[2] The local transfer entropy provides spatiotemporal profiles of information transfer, useful analytically in highlighting or filtering "hot-spots" in the information channels of the system. The local transfer entropy also facilitates close study of different forms and parameters of the averaged measure, in particular the importance of conditioning on the past history of the information destination. We also use it to investigate how the measure can condition on other information sources, in order to account for information transfer resulting from the interaction of multiple sources. Importantly, through these applications the local transfer entropy provides insights that its average cannot. We also demonstrate the manner in which information transfer combines with information storage in determining the computation of the next state of a variable.

Crucially, the local transfer entropy is shown in Sect. 4.2 to reveal particles as the dominant information transfer entities in CAs. These results have wide-ranging implications in supporting the validity of the transfer entropy itself, as well as for the real-world systems mentioned earlier. This is due to the power of CAs as model systems and the obvious analogy between particles in CAs and *coherent spatiotemporal structures* and hypothesised information transfer entities in other systems. These include known analogues of particles in physical processes such as pattern formation and solitons [4, 24], and also waves of conformational change are said to perform signalling in microtubules [8].

Finally, we demonstrate in Sects. 4.3 and 4.4 that information transfer is dependent on, though distinct from, the concept of causal information flow, which should be considered separately as a useful notion in its own right.[3] This finding is important because measures for both predictive transfer [20] and causal effect [26] have been inferred to capture information transfer in general, and measures of predictive transfer have been used to infer causality [27–30] with the two sometimes (problematically) directly equated (e.g. [31–36]). We also describe the conditions and parameter settings under which a variant of the transfer entropy converges with the information flow.

We suggest that information flow should be used first wherever possible in order to establish the set of causal information contributors for a given destination variable. Subsequently, the transfer entropy measure may be used to quantify the concept of information transfer from these causal sources (only) to the destination in order to study emergent computation in the system.

---

[2] The local transfer entropy and results of its application to CAs were first reported in [22, 23].

[3] The local information flow and results of its application to CAs were first reported in [25].

4.1 Transfer Entropy as Predictive Information Transfer

The quantitative definition of the transfer entropy has not yet been unified with the accepted specific instances of information transfer (e.g. particles in CAs). These instances are local in space and time and to be investigated require a *local* measure of information transfer. In this section, we present the transfer entropy, comment on the parameter settings required with it and contrast it to related measures. We then introduce local transfer entropy, in order to unify the apparently correct quantitative formulation of information transfer with qualitatively accepted specific instances of this concept.

4.1.1 Transfer Entropy

Mutual information $I_{Y;X}$ has been something of a de facto measure for information transfer between Y and X in complex systems science in the past (e.g. [6, 15, 37]). A major problem however is that mutual information contains no inherent *directionality*. Attempts to address this include using the previous state of the "source" variable Y and the next state of the "destination" variable X' (known as *time-lagged mutual information* $I_{Y;X'}$). However, Schreiber [20] points out that this ignores the more fundamental problem that mutual information measures the *statically* shared information between the two elements.[4]

To address these inadequacies Schreiber introduced **transfer entropy** [20] (TE), the deviation from independence (in bits) of the state transition (from the previous state to the next state) of an information destination X from the previous state of an information source Y:

$$T_{Y\rightarrow X}(k,l) = \sum_{\mathbf{w}_n} p(\mathbf{w}_n) \log_2 \frac{p(x_{n+1} \mid x_n^{(k)}, y_n^{(l)})}{p(x_{n+1} \mid x_n^{(k)})}. \tag{4.1}$$

Here n is a time index, $\mathbf{w}_n$ represents the state transition tuple $\{x_{n+1}, x_n^{(k)}, y_n^{(l)}\}$, $x_n^{(k)}$ and $y_n^{(l)}$ represent the k and l past values of x and y up to and including time n. Schreiber points out that this formulation is a truly *directional*, *dynamic* measure of information transfer, and is a generalisation of the entropy rate to more than one element to form a mutual information *rate*. That being said it remains a measure of observed (conditional) *correlation* rather than direct effect. In fact, the TE is a non-linear extension of a concept known as the "Granger causality" [38], the nomenclature for which may have added to the confusion associating information transfer and causal effect. Importantly, as an information-theoretic measure based on observational probabilities, the TE is applicable to both deterministic and stochastic systems.

The TE can be viewed as a *conditional* mutual information $I(Y^{(l)}; X' \mid X^{(k)})$ [39] (see Eq. (2.11)), casting it as the average information contained in the source about

[4] The same criticism applies to equivalent non information-theoretic definitions such as that in [19].

the next state X' of the destination that was not already contained in the destination's past $X^{(k)}$:

$$T_{Y \to X}(k, l) = I_{Y^{(l)};X' | X^{(k)}} = H_{X' | X^{(k)}} - H_{X' | X^{(k)}, Y^{(l)}}. \tag{4.2}$$

This could be interpreted (following [40] and [37]) as the diversity of state transitions in the destination minus assortative noise between those state transitions and the state of the source.

Similarly, we note that Schreiber's original description can be rephrased as the information provided by the source about the state transition in the destination. That $x_n^{(k)} \to x_{n+1}$ (or including redundant information $x_n^{(k)} \to x_{n+1}^{(k)}$) is a *state transition* is underlined in that the $x_n^{(k)}$ are *embedding vectors* [41], which capture the underlying *state* of the process.

We generalise the notation for lattice systems (such as CAs) with *spatially-ordered* agents to represent the transfer entropy from an information source X_{i-j} to an information destination X_i (i.e. information transfer across j cells to the right) as:

$$T(i, j, k, l) = \sum_{\mathbf{w}_n} p(\mathbf{w}_n) \log_2 \frac{p(x_{i,n+1} \mid x_{i,n}^{(k)}, x_{i-j,n}^{(l)})}{p(x_{i,n+1} \mid x_{i,n}^{(k)})}. \tag{4.3}$$

Again, $\mathbf{w}_n$ represents the state transition tuple $\left\{x_{i,n+1}, x_{i,n}^{(k)}, x_{i-j,n}^{(l)}\right\}$. Where the agents in the lattice system are homogeneous and the PDFs are estimated over all agents, it is appropriate to average over all agents i also to obtain $T(j, k, l)$.

In most cases we consider systems where only the previous state of the source is a causal contributor to the destination (e.g. CAs). For these systems we use $l = 1$ and simply drop it from the notation $T_{Y \to X}(k)$ from here on.[5] Note however that all of the following equations in this chapter may be formed with $l > 1$ if appropriate.

The TE has already been widely used to analyse information transfer, for example in interactions in the stock market [42], food webs [43], EEG signals [44], and biochemicals [45]. However, there are a number of features and perspectives of this measure which require addressing for our study of distributed computation. These include the appropriate past history lengths k to use, and how to include interactions of multiple sources. We consider these in the following sections (including the remainder of Sect. 4.1).

4.1.1.1 Conditioning on an Infinite History Length

It is important to note that the destination's own historical values can indirectly influence it via the source or other neighbours: this may be mistaken as an independent

[5] This aligns with Sect. 4.4.2 which demonstrates that in accurately capturing information transfer we should only examine causal sources here.

transfer of information from the source here. This can be facilitated in systems with *cyclic* causal paths (e.g. the *bidirectional* causal links in CAs). Such self-influence is a *non-travelling* form of information (in the same way as standing waves are to energy); it is essentially static and can be viewed as the trivial part of information transfer. In the context of distributed computation, it is recognisable as the *active information storage* and this is made explicit in:

$$T_{Y\to X}(k) = I_{Y;X'|X^{(k)}} = I_{X';\{Y,X^{(k)}\}} - I_{X';X^{(k)}}, \quad (4.4)$$

$$= I_{X';\{Y,X^{(k)}\}} - A_X(k). \quad (4.5)$$

Conditioning on the destination's history $x_{i,n}^{(k)}$ serves to eliminate the active information storage from the transfer entropy measurement. Yet any self-influence transmitted prior to these k values will not be eliminated; we generalise comments on the entropy rate in [20] to suggest that taking the asymptote $k \to \infty$ is most correct for agents displaying non-Markovian dynamics (when considering their time-series in isolation). Just as the excess entropy and active information storage require $k \to \infty$ to *capture* all information storage, accurate measurement of the TE requires $k \to \infty$ to *eliminate* all information storage from being mistaken as information transfer. **The most correct form of the transfer entropy to measure information transfer is then**:

$$T_{Y\to X} = \lim_{k\to\infty} \sum_{\mathbf{w}_n} p(\mathbf{w}_n) \log_2 \frac{p(x_{n+1} \mid x_n^{(k)}, y_n)}{p(x_{n+1} \mid x_n^{(k)})}. \quad (4.6)$$

Computation at this limit is not feasible in general, so we retain $T_{Y\to X}(k)$ for estimation with finite-k though of course recommend the use of as large a k as is practical.[6] For the lattice form in Eq. (4.3) we write $T(i, j)$ and $T(j)$ in this limit. This recommendation of using as long a value of k as possible is important because in most applications (e.g. to characterise information flow in sensorimotor networks [14] and with respect to information closure [46]) the TE has been applied with a default value of $k = 1$.

4.1.1.2 Related Information Transfer Style Measures

We note the similar *information current* [47], which measures changes in *spatial* information and does so on a local scale in space and time. This measure is interpretable as an information contribution between the right and left segments of a system only for those exhibiting deterministic mechanics. Furthermore, in considering spatial information it is only defined for lattice systems, where it either measures information contribution from the right side of the system to the left (or vice-versa) but not from an arbitrary source to an arbitrary destination. As such, it remains out

6 See the discussion regarding practical values of k being limited by the number of available observations for the active information storage in Sect. 3.3.1.

of scope for our investigation of information transfer between a specific source and destination in general multivariate systems.

We also note the close relationship between the transfer entropy and Massey's directed information [48]; in Appendix C we describe the subtle differences between the two measures and why the TE is the most appropriate measure of information transfer in *distributed computation*.

Furthermore, we note that damage spreading or indeed the uncertainty in perturbation avalanche or cascade size have been suggested [11] and used [49] to infer information transfer. While information transfer is likely to be an integral part of cascade dynamics, measuring it as such is problem-specific, cannot be used on observations alone, and cannot resolve the contribution of multiple sources. We will investigate the relationship of information transfer (using the TE) to such cascades in Sect. 6.2.

4.1.2 Local Transfer Entropy

Using the technique described in Sect. 2.2.2,[7] we observe that the TE is an average (or *expectation value*) of a **local transfer entropy** at each observation n, i.e.:

$$T_{Y\to X} = \langle t_{Y\to X}(n+1) \rangle, \tag{4.7}$$

$$t_{Y\to X}(n+1) = \lim_{k\to\infty} t_{Y\to X}(n+1,k), \tag{4.8}$$

$$T_{Y\to X}(k) = \langle t_{Y\to X}(n+1,k) \rangle, \tag{4.9}$$

$$t_{Y\to X}(n+1,k) = \log_2 \frac{p(x_{n+1} \mid x_n^{(k)}, y_n)}{p(x_{n+1} \mid x_n^{(k)})}. \tag{4.10}$$

The local transfer entropy quantifies the information contained in the source y_n about the next state of the destination x_{n+1} at time step $n+1$, that was not contained in the past of the destination $x_n^{(k)}$. The measure is *local* in that it is defined at each time n for each destination element X in the system and each causal information source Y of the destination. For example, the local transfer entropies at time step $n+1$ from two different causal sources Y_1 and Y_2 to a destination X are indicated in Fig. 4.1, along with the active information storage for X.

The local TE may also be expressed as a local conditional mutual information:

$$t_{Y\to X}(n+1,k) = i(y_n; x_{n+1} \mid x_n^{(k)}). \tag{4.11}$$

Similarly, we can write the local time-lagged mutual information as $i(y_n; x_{n+1})$.

For lattice systems such as CAs, we represent the local transfer entropy from X_{i-j} to X_i (i.e. across j cells to the right) at time $n+1$ as $t(i,j,n+1)$:

[7] As discussed in Sect. 2.2.2, a "local" transfer entropy has been presented in [30]; however this uses a sliding-window technique which does not measure the information transfer involved in the computation at a *single specific time step*.

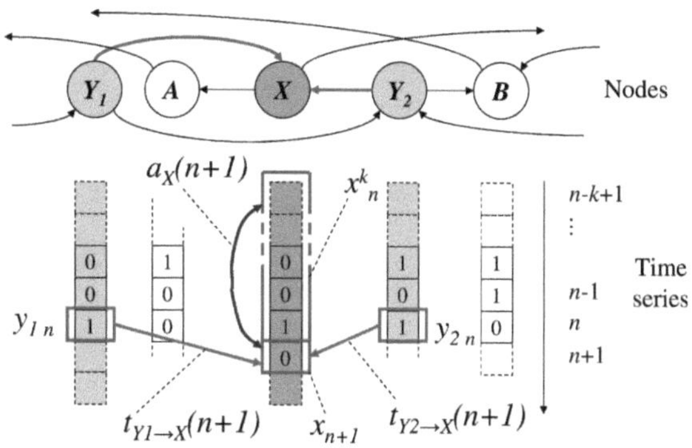

Fig. 4.1 Information dynamics in a multivariate system. For node X, this figure displays the local active information $a_X(n+1,k)$ and the local transfer entropies $t_{Y_1\rightarrow X}(n+1)$ and $t_{Y_2\rightarrow X}(n+1)$ from each of the causal information sources $\mathbf{V}_X = \{Y_1, Y_2\}$ at time $n+1$. (NB: This figure was first published in [50])

$$T(i,j) = \langle t(i,j,n+1)\rangle_n , \tag{4.12}$$

$$t(i,j,n+1) = \lim_{k\rightarrow\infty} t(i,j,n+1,k), \tag{4.13}$$

$$T(i,j,k) = \langle t(i,j,n+1,k)\rangle_n , \tag{4.14}$$

$$t(i,j,n+1,k) = \log_2 \frac{p(x_{i,n+1} \mid x_{i,n}^{(k)}, x_{i-j,n})}{p(x_{i,n+1} \mid x_{i,n}^{(k)})}, \tag{4.15}$$

$$= i(x_{i-j,n}; x_{i,n+1} \mid x_{i,n}^{(k)}). \tag{4.16}$$

We can also write $T(j,k) = \langle t(i,j,k,n+1)\rangle_{i,n}$ for systems of homogeneous agents. Similarly, the local (time-lagged) mutual information can be represented as: $i(i,j,n+1) = i(x_{i-j,n}; x_{i,n+1})$.

This local TE $t(i,j,n+1,k)$ to agent X_i from X_{i-j} at time $n+1$ in a lattice system is illustrated in Fig. 4.2a. $t(i,j,n,k)$ is defined for every spatiotemporal destination (i,n), forming a *spatiotemporal profile* for every information channel or direction j. Sensible values for j correspond to causal information sources, i.e. for CAs, sources within the cell range $|j| \leq r$.[8]

4.1.3 Apparent, Conditional and Complete Transfer Entropy

As per the active information storage, the average transfer entropy $T_{Y\rightarrow X}(k)$ is always positive but is bounded above by the information capacity of a single state of the destination. For a discrete system with b possible states this is $\log_2 b$ bits. As a conditional mutual information, it can be either larger *or* smaller than the corresponding

[8] We demonstrate in Sect. 4.4.2 that the transfer entropy is interpretable as information transfer for these sources only.

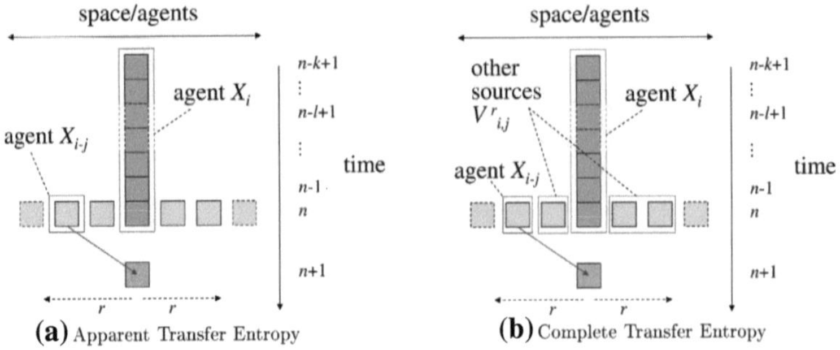

Fig. 4.2 Transfer entropy in CAs (after [23, 25]). **a** Apparent TE $t(i, j, n+1, k)$: information contained in the source cell X_{i-j} about the next state of the destination cell X_i at time $n+1$ that was not contained in the destination's past. **b** Complete TE $t^c(i, j, n+1, k)$: information contained in the source cell X_{i-1} about the next state of the destination cell X_i at time $n+1$ that was not contained in *either* the destination's past *or* all of the other causal information sources $\mathbf{V}^r_{i,j}$. For CAs these causal sources are within the cell range r

mutual information [51]. The *local* TE however is not constrained so long as it averages into this range: it can be greater than $\log_2 b$ for a large local information transfer, and can also in fact be measured to be negative. This is a similar situation to that which we saw regarding the local information storage in Sects. 3.1.2 and 3.2.1. Local transfer entropy is negative where (in the context of the history of the destination) the probability of observing the actual next state of the destination given the value of the source $p(x_{i,n+1} \mid x^{(k)}_{i,n}, x_{i-j,n})$, is lower than that of observing that actual next state independently of the source $p(x_{i,n+1} \mid x^{(k)}_{i,n})$. In this case, the source element is actually *misinformative* or misleading about the state transition of the destination. It is possible for the source to be misleading where other causal information sources influence the destination, or in a stochastic system.

It is also possible for the measure to attribute an effect to one source that was in reality due to another source. As such, we label the previously described measures as **apparent transfer entropy** to explicitly indicate they measure an apparent effect without taking other sources into account.

In contrast the transfer entropy may be conditioned on other possible causal information sources Z, to eliminate their influence from being attributed to the source in question Y [20]. We term this a **conditional transfer entropy**, written on average and locally as:

$$T_{Y\to X|Z} = \langle t_{Y\to X|Z}(n+1)\rangle_n, \tag{4.17}$$

$$t_{Y\to X|Z}(n+1) = \lim_{k\to\infty} t_{Y\to X|Z}(n+1,k), \tag{4.18}$$

$$T_{Y\to X|Z}(k) = \left\langle t_{Y\to X|Z}(n+1,k)\right\rangle_n, \tag{4.19}$$

$$t_{Y\to X|Z}(n+1,k) = \log_2 \frac{p(x_{n+1} \mid x^{(k)}_n, y_n, z_n)}{p(x_{n+1} \mid x^{(k)}_n, z_n)}, \tag{4.20}$$

$$= i(y_n; x_{n+1} \mid x^{(k)}_n, z_n). \tag{4.21}$$

The conditional transfer entropy is the average information contained in the source Y about the next state of the destination X' that was not contained in the destination's past *or* in the source(s) Z. Elsewhere, this quantity has been termed the *partial* transfer entropy [52] and considered in [30].

Were one seeking to eliminate the influence of other sources, then ideally one would condition on *all* sources Z in the set of parents or causal information contributors $\mathbf{V}_X$ to X, except for the source Y and X itself. We denote this set as $\mathbf{V}_X^Y$:

$$\mathbf{V}_X^Y = \{Z \in \mathbf{V}_X, Z \neq Y, Z \neq X\}, \tag{4.22}$$

$$= \mathbf{V}_X \setminus X, Y. \tag{4.23}$$

For example, in Fig. 4.1 we have $\mathbf{V}_X = \{X, Y_1, Y_2\}$, $\mathbf{V}_X^{Y_1} = \{Y_2\}$ and $\mathbf{V}_X^{Y_2} = \{Y_1\}$. At time step n, this set has joint state $\mathbf{v}_{x,n}^y$:

$$\mathbf{v}_{x,n}^y = \left\{ z_n \mid \forall Z \in \mathbf{V}_X^Y \right\}. \tag{4.24}$$

We then introduce the **complete transfer entropy** as the average information contained in the source about the next state of the destination that was not contained in the destination's past *or* in other causal information sources $\mathbf{v}_{x,n}^y$. We write the average measure, and the **local complete transfer entropy**, as:

$$T_{Y\to X}^c = \left\langle t_{Y\to X}^c(n+1) \right\rangle_n, \tag{4.25}$$

$$t_{Y\to X}^c(n+1) = \lim_{k\to\infty} t_{Y\to X}^c(n+1,k), \tag{4.26}$$

$$T_{Y\to X}^c(k) = \left\langle t_{Y\to X}^c(n+1,k) \right\rangle_n, \tag{4.27}$$

$$t_{Y\to X}^c(n+1,k) = \log_2 \frac{p(x_{n+1} \mid x_n^{(k)}, y_n, \mathbf{v}_{x,n}^y)}{p(x_{n+1} \mid x_n^{(k)}, \mathbf{v}_{x,n}^y)}, \tag{4.28}$$

$$= i(y_n; x_{n+1} \mid x_n^{(k)}, \mathbf{v}_{x,n}^y). \tag{4.29}$$

Again, notice that the expressions are only completely correct as information transfers in the limit $k \to \infty$.

Our language above in describing the conditioning out of the influence of other sources may have indicated that the conditional and complete measures will give smaller values than the apparent TE. However, the extra conditioning in these conditional mutual information values can make *either* larger or smaller than the original measure (see Sect. 2.2.1). As an illustration, consider sources Y_1 and Y_2 acting on X in an exclusive-OR (XOR) operation. Assuming maximum entropy sources, one needs knowledge of both Y_1 and Y_2 in order to predict X. As such, the apparent TE for either source will give 0 bits, while the complete TE for either will give 1 bit in full prediction of X. This example illustrates that the apparent TE is directed towards capturing **single-source transfer** only, or the **coherent transfer** effect of one source on a destination. In contrast, the complete TE conditions out transfer from other sources, but also *additionally* captures **interaction-based transfer**, i.e. the transfer resulting from the interaction of the given source with other sources.

We also introduce specific notation for these quantities in CAs. There, the set of causal information contributors to X_i is the neighbourhood $\mathbf{V}_{i,r}$ of X_i within the range r:

$$\mathbf{V}_{i,r} = \left\{X_{i-q} \mid \forall q : -r \leq q \leq +r\right\}. \tag{4.30}$$

At time step n, this set has joint value $\mathbf{v}_{i,r,n}$:

$$\mathbf{v}_{i,r,n} = \left\{x_{i-q,n} \mid \forall q : -r \leq q \leq +r\right\}. \tag{4.31}$$

For the complete TE we condition on this set except for the source X_{i-j} and X_i itself, $\mathbf{V}^j_{i,r}$:

$$\mathbf{V}^j_{i,r} = \left\{X_{i-q} \mid \forall q : -r \leq q \leq +r, q \neq j, q \neq 0\right\}. \tag{4.32}$$

At time step n this set has joint value $\mathbf{v}^j_{i,r,n}$:

$$\mathbf{v}^j_{i,r,n} = \left\{x_{i-q,n} \mid \forall q : -r \leq q \leq +r, q \neq j, q \neq 0\right\}. \tag{4.33}$$

We can then provide the following expressions for the complete transfer entropy in CAs on average and in local notation:

$$T^c(i, j) = \left\langle t^c(i, j, n+1)\right\rangle_n, \tag{4.34}$$

$$t^c(i, j, n+1) = \lim_{k \to \infty} t^c(i, j, n+1, k), \tag{4.35}$$

$$T^c(i, j, k) = \left\langle t^c(i, j, n+1, k)\right\rangle_n, \tag{4.36}$$

$$t^c(i, j, n+1, k) = \log_2 \frac{p\left(x_{i,n+1} \mid x^{(k)}_{i,n}, x_{i-j,n}, \mathbf{v}^j_{i,r,n}\right)}{p\left(x_{i,n+1} \mid x^{(k)}_{i,n}, \mathbf{v}^j_{i,r,n}\right)}, \tag{4.37}$$

$$= i(x_{i-j,n}; x_{i,n+1} \mid x^{(k)}_{i,n}, \mathbf{v}^j_{i,r,n}). \tag{4.38}$$

Additionally, in homogeneous systems such as CAs we can average over all agents also to obtain $T^c(j) = \langle t^c(i, j, n+1)\rangle_{i,n}$ and $T^c(j, k) = \langle t^c(i, j, n+1, k)\rangle_{i,n}$. The quantity $t^c(i, j, n+1, k)$ is displayed in Fig. 4.2b.

In deterministic systems (e.g. CAs), conditioning on all causal source renders $t^c_{Y \to X}(n+1, k) \geq 0$. This is because the only possible observed value of x_{n+1} as determined by $\{y_n, x^{(k)}_n, \mathbf{v}^y_{x,n}\}$ has the numerator of the log term in Eq. (4.26) as $p\left(x_{n+1} \mid x^{(k)}_n, y_n, \mathbf{v}^y_{x,n}\right) = 1$ and a denominator less than or equal to this.

Note that only causal sources have non-zero complete TE. Where it is measured for a non-causal source, the denominator of the log term in Eq. (4.26) is

$p\left(x_{n+1} \mid x_n^{(k)}, \mathbf{v}_{x,n}^{y}\right) = 1$ since the next state of the destination is fully determined without that source. An implication here is that in systems such as CAs where only immediately previous source values can be causal information contributors, using $l > 1$ will not alter the value of the complete TE because there is no more information to add to the source.

In parallel, one may naively suggest that taking $k > 1$ with the complete TE in such systems is redundant. It is true that the historical values of the destination cannot add information about the next state once all causal sources have been examined. However, they are being not examined as information sources here. They are used to *condition away* any (stored) information that they share (or have correlated) with the source in question, and for this role they are not redundant.

4.1.4 Total Information Composition and Collective Information Transfer

In Sect. 3.2.2 we demonstrated that the information required to predict the next state of any agent (the local entropy) can be decomposed into the amount predictable from its past (the local active information storage) and the remaining uncertainty after examining this memory (the local temporal entropy rate). Here we show how the local temporal entropy rate may subsequently be decomposed as a sum of transfer entropy terms (plus intrinsic uncertainty).[9]

The temporal entropy rate is composed of information that was not in the destination's past but is added jointly to the next state X' of the destination by the causal sources $\mathbf{V}_X$ ($I_{\mathbf{V}_X;X'|X^{(k)}}$), and the remaining **intrinsic uncertainty** U_X[10]:

$$H_{\mu X}(k) = I_{\mathbf{V}_X;X'|X^{(k)}} + U_X(k), \tag{4.39}$$

$$U_X(k) = H_{X'|\mathbf{V}_X,X^{(k)}}. \tag{4.40}$$

Note that the validity of the relation is independent of k. The $I_{\mathbf{V}_X;X'|X^{(k)}}$ term is clearly a transfer entropy, representing the joint information transfer from all of the

[9] Similarly motivated are the decompositions of the total information for a destination presented in [53] and [54]. The approach in [53] however decomposes the total information in terms of a sum of mutual information terms of increasing size, rather than incrementally conditioned mutual information terms. The approach in [54] does use incrementally conditioned mutual information terms, though again this is merely as a means to infer the contribution to the total information from terms of increasing size or order k. Our presentation is distinct from both of these in considering the destination's history as a single source, regardless of whether these states are causal contributors or not. Establishing the context of the past is critical for an understanding of distributed computation. Also, our approach is distinct in considering the decomposition on a local scale.

[10] Note that any dependence of U_X on k is due to statistical fluctuations rather than a causal effect. (Of course, this is for k large enough to include all direct causal contributors in the past of the destination).

causal sources $\mathbf{V}_X$ to X.[11] As such, we call this the **collective transfer entropy**:

$$T_X(k) = I_{\mathbf{V}_X;X'|X^{(k)}}, \tag{4.41}$$
$$T_X = \lim_{k\to\infty} T_X(k). \tag{4.42}$$

For deterministic systems (e.g. CAs) $U_X(k) = 0$, so the temporal entropy rate is equivalent to the collective transfer entropy: $H_{\mu X}(k) = T_X(k)$. This analytically explains the assertion in Sect. 3.3.7 that the local temporal entropy rate represents a collective information transfer in deterministic systems. It also underpins why the local temporal entropy rate $h_\mu(i, n, k)$ was shown to be a useful filter for all moving particles in Sect. 3.3.7, since these particles are strongly predictable from the causal neighbours $\mathbf{V}_X$ in the context of the past $X^{(k)}$.

The collective TE can then be decomposed itself by *incrementally* taking account of the contribution of each causal information source (i.e. by following the chain rule [55] for it as a conditional mutual information). Let us consider an *arbitrary* ordering (using index g) of the causal sources in $\{\mathbf{V}_X \setminus X\}$: $Z_1, Z_2, ..., Z_G$. We can then write an arbitrarily ordered subset of $g - 1$ sources as:

$$\mathbf{V}_X^g = \{Z_c \mid \forall c : 1 \le c < g\}, \tag{4.43}$$

and then make the decomposition:

$$T_X(k) = \sum_g I(Z_g; X' \mid X^{(k)}, \mathbf{V}_X^g). \tag{4.44}$$

We have a sum of *incrementally conditional mutual information* terms: each term is the information added by the given source Z_g that was not contained either in the past of the destination or in the previously inspected sources $\mathbf{V}_X^g$. Each term is a transfer entropy itself, and if we expand this sum:

$$\begin{aligned} T_X(k) =& I(Z_1; X' \mid X^{(k)}) + I(Z_2; X' \mid X^{(k)}, Z_1) + \\ & I(Z_3; X' \mid X^{(k)}, Z_1, Z_2) + \cdots + I(Z_G; X' \mid X^{(k)}, \mathbf{V}_X^G), \end{aligned} \tag{4.45}$$

we see that the first term is the apparent TE from source Z_1, the last term is the complete TE from source Z_G (since all other causal sources are conditioned on), and the intermediate terms are conditional transfer entropies (see Eq. (4.19)). The collective transfer entropy captures (while accounting for redundancies) all transfers from the sources to the destination, incorporating both single-source and interaction-based transfers. Importantly, it is not a simple sum of the apparent TE from each source, nor the sum of the complete TE from each source.

---

[11] Conceptually, the collective transfer entropy is intended to consider all causal sources to X apart from X itself. Mathematically though, we do not need to exclude X from $\mathbf{V}_X$ here since it is conditioned out in $X^{(k)}$ (so long as $k \ge 1$).

Finally, we can combine Eqs. (3.22), (4.39) and (4.44) together:

$$H_{X'} = A_X(k) + T_X(k) + U_X(k), \tag{4.46}$$
$$H_{X'} = A_X(k) + \sum_g I(Z_g; X' \mid X^{(k)}, \mathbf{V}_X^g) + U_X(k). \tag{4.47}$$

This explicitly demonstrates that the information required to predict the next state of a destination is the sum of:

- the information gained from the past of the destination (i.e. the active information storage); plus
- the collective information transfer from all causal sources (as a sum of incrementally conditioned transfer entropies); plus
- any remaining intrinsic uncertainty in the destination.

As previously commented, these equations are correct for any value of k, however we note that only the use of $k \to \infty$ will show the true separation between information storage and transfer in contributing to the computation of the next state of the destination. This again underlines the importance of the context of past history (in establishing the underlying state of the destination) for our perspective of distributed computation.

Of course, it is trivial to write all of the above in this limit $k \to \infty$, as well as in lattice notation for CAs, and with local measures of each. For example, we denote finite-k estimates of the **local collective transfer entropy** using $t_X(n+1,k)$:

$$T_X = \langle t_X(n+1) \rangle_n , \tag{4.48}$$
$$t_X(n+1) = \lim_{k\to\infty} t_X(n+1,k), \tag{4.49}$$
$$T_X(k) = \langle t_X(n+1,k) \rangle_n , \tag{4.50}$$
$$t_X(n+1,k) = \log_2 \frac{p(x_{n+1} \mid x_n^{(k)}, \mathbf{v}_{x,n})}{p(x_{n+1} \mid x_n^{(k)})}, \tag{4.51}$$
$$= i(\mathbf{v}_{x,n}; x_{n+1} \mid x_n^{(k)}), \tag{4.52}$$

where $\mathbf{v}_{x,n}$ is the joint value of $\mathbf{V}_X$ at time n. Also, where at time n the joint value of $\mathbf{V}_X^g$ is $\mathbf{v}_{x,n}^g$, the value of Z_g is $z_{g,n}$, and the value of $U_X(k)$ is $u_X(n,k)$, we can write Eqs. (4.46) and (4.47) on a local scale in time as:

$$h_X(n+1) = a_X(n+1,k) + t_X(n+1,k) + u_X(n+1,k), \tag{4.53}$$
$$h_X(n+1) = a_X(n+1,k) + \sum_g i(z_{g,n}; x_{n+1} \mid x_n^{(k)}, \mathbf{v}_{x,n}^g) + u_X(n+1,k). \tag{4.54}$$

We can similarly define these quantities in lattice systems, the finite-k estimate of the local collective TE for example is:

$$t(i,n+1,k) = i(\mathbf{v}_{i,r,n}; x_{i,n+1} \mid x_{i,n}^{(k)}). \tag{4.55}$$

We emphasise also that the order of considering each causal source here is arbitrary, though in lattice systems spatial ordering is an obvious candidate. As a demonstrative example, in ECAs in local lattice notation we have:

$$h(i,n+1) = a(i,n+1,k) + t(i,j=-1,n+1,k) + t^c(i,j=1,n+1,k), \tag{4.56}$$

or in reverse source ordering:

$$h(i,n+1) = a(i,n+1,k) + t(i,j=1,n+1,k) + t^c(i,j=-1,n+1,k). \tag{4.57}$$

In the above, the different forms of the TE as information transfer from causal sources can be seen to characterise important components of the total information at the destination. Neither the apparent, conditional nor complete forms are more correct than the others. They are all measures of information transfer, but answer subtly different questions regarding how the destination computes its next state. As we see here, the answers they provide are complementary. Indeed, we see that *together* they quantify the information that is *physically added* to the destination in a distributed computation, in contrast to the stored information which is considered to already be present in the destination. The context of the past history of the destination is critical to the perspective of distributed computation; it is only the use of this context that allows us to properly separate information storage and transfer here.

Importantly, note that no non-causal information sources appear in the sum of information transfer terms contributing to the total information at the destination. This is not the case for the past history of the destination, since from the perspective of computation we consider its contribution as information storage even if it is not a causal contributor (as described in Sect. 3.2.2).

4.1.4.1 Excess Entropy in Terms of Collective Transfer Entropy

Finally, we present a relationship between the collective transfer entropy and the excess entropy. In a similar fashion to Eq. (3.30) in Sect. 3.2.2.2, we substitute Eqs. (4.39) and (4.41) into Eq. (2.18), getting:

$$E_X = \sum_{k=0}^{\infty} [T_X(k) - T_X]. \tag{4.58}$$

This expression shows that the excess entropy is the sum of overestimates of the collective information transfer at each finite history length k. The overestimates are due to the observer not accounting for storage structure beyond the history length k.

As per Sect. 3.2.2.2, we could display this relationship graphically in a similar fashion to the plot for the excess entropy in terms of entropy rates in [56]. We have not included this plot here however, as it would appear identical to the plot of $H_{\mu X}(k)$ in Fig. 3 of [56] except that the curve and asymptote would be reduced by $U_X(k)$ (since $T_X(k) = H_{\mu X}(k) - U_X(k)$).

4.2 Local Information Transfer in Cellular Automata

Local complete and apparent transfer entropy were applied with $k = 16$ to the same sample ECA runs where information storage was analysed in Sect. 3.3; i.e. for rules 54 (Fig. 3.4), 110 (Fig. 3.5), ϕ_{par} (Fig. 3.6), 18 (Fig. 3.9), 22 (Fig. 3.10) and 30 (Fig. 3.11). As previously discussed in Sect. 3.3.7 these figures also include the local temporal entropy rate $h_\mu(i, n+1, k)$, which is equal to the local collective TE $t(i, n+1, k)$ in these deterministic systems.
In this section, we discuss the key results here:

- Mutual information and indeed TE with history length $k = 1$ are inadequate measures of information transfer (Sect. 4.2.1).
- When used with appropriate history lengths, the transfer entropy measures much greater information transfer in particles than background domains in Sect. 4.2.2. In aligning with our qualitative notion of the concept, it is thus shown to correctly measure information transfer in distributed computation.
- While the particles are the *dominant* information transfer entities, they are not unique such entities: in Sect. 4.2.3 we describe why transfer is also found throughout the domain region.
- We also explore the complementary relationship between the apparent and complete TE in Sect. 4.2.4, including the important situation where the local apparent values become misinformative.

4.2.1 Inadequate Measures for Information Transfer

As base cases we measured for rule 110 the (time-lagged) local mutual information $i(i, j, n)$, and local apparent and complete transfer entropies with the default value of $k = 1$: $t(i, j, n, k = 1)$ and $t^c(i, j, n, k = 1)$. The comparison to the local mutual information is analogous to that with averaged measures in [20], yet the local profiles yield a more detailed contrast here than averages do. Note that $k = 1$ is the only value used in the original presentation of TE in [20] (in less coupled systems) and in most later applications, e.g. [14, 26, 46].

The local profiles generated with $j = 1$ (i.e. one cell to the right per unit time) for these base cases are shown in Fig. 4.3. These measures are unable however to distinguish gliders from the background here with any more clarity than the raw CA plot itself. They were also unsuccessful with other values of j and with other CA rules. As hypothesised earlier, a true measure of information transfer should align

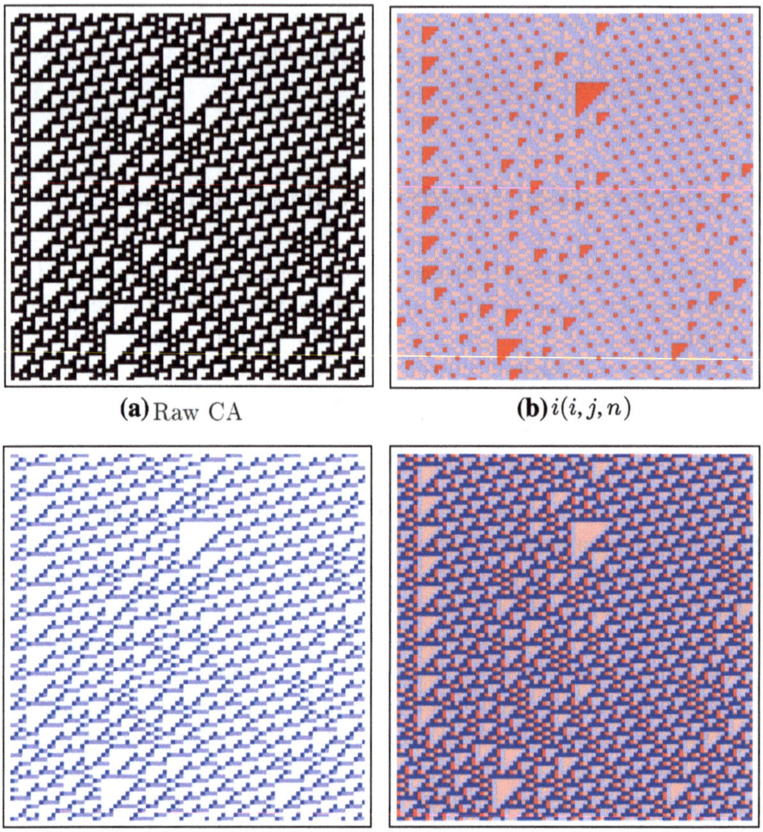

Fig. 4.3 Base comparison measures incapable of quantifying local information transfer (*one cell to the right*) in rule 110 (86 time steps displayed for 86 cells): **b** Local (*time-lagged*) mutual information $i(i, j = 1, n)$, max. 0.48 bits, min. −1.06 bits; **c** Local complete transfer entropy $t^c(i, j = 1, n, k = 1)$, max. 1.28 bits, min. 0.00 bits; **d** Local apparent transfer entropy $t(i, j = 1, n, k = 1)$, max. 0.67 bits, min. −0.61 bits. (NB: This figure is reprinted from [23] with permission of the American Physical Society)

with the qualitative notion of gliders as information transfer. This provides explicit demonstration that **the MI and TE with $k = 1$ are inadequate as measures of information transfer in distributed computation**. The TE measures are inadequate with $k = 1$ because they do not condition on a long enough extent of the past history of the destination. As suggested in Sect. 4.1.1.1, they do not therefore eliminate large amounts of what is actually information storage here.

4.2.2 *Particles as Dominant, Coherent Information Transfer Structures*

With sufficiently large values of k, we eliminate storage from the TE measurement and find vanishing transfer in the domain regions (we discuss the effect of the history length in Sect. 4.2.3 as well as non-zero levels of transfer in the domains). We then find that the transfer entropy does indeed meet our expectation of identifying particles (including gliders and domain walls) as strong positive information transfer as compared to the domain regions. (See the plots with $k = 16$ for rule 54 in Fig. 3.4, rule 110 in Fig. 3.5, rule ϕ in Fig. 3.6 and rule 18 in Fig. 3.9). More precisely, they can be distinguished from the background because the information transfer in their direction of motion is much stronger than the information storage at the same points (i.e. $t(i, j, n) > a(i, n)$). This is the case for **both** the local apparent and complete TE, and for gliders as well as domain walls. Importantly, the particles are measured as information transfer in the direction of their macroscopic motion, as expected from such measures. **These results provide quantitative evidence that particles are the dominant information transfer entities in CAs**. They also unify the theoretical appeal of the transfer entropy with an ability to identify recognised local information transfer structures.

Simply relying on the average TE values does not tell us whether gliders or particles exist in a given CA though. For instance, the profiles of rule 22 in Fig. 3.10 and rule 30 in Fig. 3.11 contain much information transfer: e.g. the average value $T(j = 1, k = 16) = 0.19$ bits for rule 22 is greater than for rule 110 at 0.07 bits. However only the local TE profiles reveal the absence of particles for rules 22 and 30 (similar to the local information storage results in Sect. 3.3.8). Furthermore, the averages do not tell us whether the length k is appropriate either (as found for the active information storage in Sect. 3.3.1, and further discussed in Sect. 4.2.3). **It is only the local TE values that answer these questions and locates these *coherent information transfer structures***. They are coherent both in terms of being composed of contiguous spatiotemporal points with large transfer, and in that the apparent TE reveals the large coherent transfer from a single source to the destination.

As an example, at the "x" marks in Fig. 3.8 which denote parts of the right-moving γ^+ gliders in rule 54, we have $p(x_{i,n+1} \mid x_{i,n}^{(k=16)}, x_{i-1,n}) = 1.00$ and $p(x_{i,n+1} \mid x_{i,n}^{(k=16)}) = 0.25$. There is a strong information transfer of $t(i, j = 1, n, k = 16) = 2.02$ bits here because the source (in the glider) added a significant amount of information to the destination in comparison to its past (in the domain). As described in Sect. 3.3.3 these points identified as part of the glider from our temporal perspective of computation are different to those identified from a perspective of spatial pattern recognition (e.g. in [57]).

We emphasise corresponding results *confirming domain walls as strong information transfer entities* in each channel for rule 18 Fig. 3.9. As the domain wall moves, it transfers information *in the direction of movement*. A full picture encompassing both directions of movement is given by the local entropy rate profile in Fig. 3.9d. As mentioned in Sect. 4.1.4, these are equivalent to the collective TE profiles $t(i, n, k)$ in deterministic systems, and so capture transfer from all causal sources.

For ϕ_{par} we confirm the role of the gliders as information transfer entities in the human understandable computation of density classification, and demonstrate information transfer across multiple units of space per unit time step for fast-moving gliders in Fig. 3.6f.

Finally, we note that the highlighting of structure by local transfer entropy is similar to results from other methods of *filtering* for emergent structure in CAs [6, 58–62]. This is because (as described above) these particles embody large amounts of information transfer in their direction of motion, allowing them to be highlighted or filtered by measuring information transfer in that direction of motion. In particular, we emphasise the local temporal entropy rate (in capturing *collective* TE here) provides useful filtering for *all* moving emergent particle structures, rather than just those moving in a particular direction. Moving particles are highlighted by other filtering techniques because they also happen to embody large values of the measures used by those techniques (e.g. large local statistical complexity [58]). The distinguishing features of local transfer entropy filtering compared to these other techniques are discussed in the summary of this chapter. Worth mentioning here is that much less of the gliders are highlighted than for other CA filtering techniques. Also, the larger values of local TE are concentrated around the leading time-edges of the gliders. In alignment with Sect. 3.3.4, this suggests that the leading glider edges determine much of the following dynamics which are then mainly information storage processes.

4.2.3 Ambient Transfer in Backgrounds Domains

As discussed in Sect. 4.1.1.1 and inferred in Sects. 4.2.1 and 4.2.2, a sufficient history length k is required to eliminate information storage in the periodic domain regions from being measured as information transfer here. For example, we plot $T(j, k)$ as a function of k for rule 110 in Fig. 4.4 (in analogy to Fig. 3.3 for the information storage). This plot shows that much information storage is eliminated with $k \geq 7$.[12]

To understand why this is the case, we revisit the argument of Sect. 3.3.2. Were the domain region infinite (i.e. involving only coupled periodic processes) with temporal period p, then using $k >= p - 1$ would measure $t(i, n, k = p - 1) = 0$ as expected.

[12] We also note that Fig. 4.4 shows the average complete transfer entropies decrease with k: an increase is impossible because we condition out more of the information that appears to come from the source. The average apparent transfer entropy can show increases with k however; this is possible with a three-term entropy [51] where other information sources are not taken into account. None of these reach a limiting value for the extent of k measured, suggesting again that as large a k as is practical should be used. Also, while the average apparent and complete measures may appear to be converging to a similar value in each channel, this belies their important distinctions which are discussed in Sect. 4.2.4.

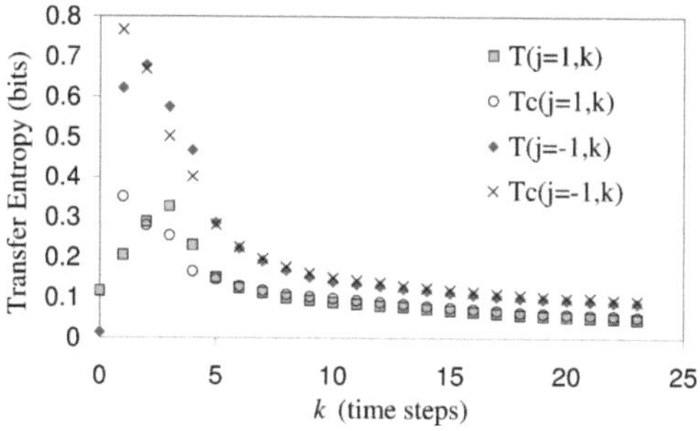

Fig. 4.4 Average transfer entropies versus history length k, plotted for complete and apparent transfer entropies in channels $j = 1$ and -1 in ECA rule 110. (NB: This figure is reprinted from [23] with permission of the American Physical Society)

This is because the processes are then completely predictable from their past.[13] **For a destination time-series with *punctuated* periodic sequences** (i.e. like the periodic domains here), **setting $k = p - 1$ will eliminate the information storage related to the period of these sequences and is a *useful minimum value*** (hence why $k \geq 7$ appears useful for rule 110 from Fig. 4.4). With this minimum setting, the gliders are then shown to be dominant information transfer entities in comparison to the domains (as per the results for $a(i, n, k = 7)$ for rule 110 in Fig. 3.7c). However, this setting will not eliminate any longer-range correlations (e.g. one glider type may be more prevalent after encountering another type) so ideally the limit $k \to \infty$ should be taken.[14]

Importantly, **there is another small but non-zero information transfer component in the periodic background domains that is not a finite-k effect**. Since domains are periodic processes occasionally punctuated by gliders, the next state of a cell there is not always completely determined by its periodic history. There is scope for the neighbouring sources to *add* information about the next state of that destination, effectively indicating whether a glider is incoming or not. So a non-zero transfer within the domain could represent the *absence* of a glider coming from the source cell (i.e. that the domain shall continue). For example, in Fig. 3.8 for rule 54 consider the points in the periodic domain two cells to the right and two time steps back (up) from those marked “x”. These points have the same history as those marked “x” which as part of the γ^+ glider have high transfer to the right ($j = 1$); their neighbourhood (excluding the source on the left) is also the same. Here, we compute $p(x_{i,n+1} \mid x_{i,n}^{(k=16)}, x_{i-1,n}) = 0.87$ and $p(x_{i,n+1} \mid x_{i,n}^{(k=6)}) = 0.75$: the probability of observing the actual next state of the destination becomes slightly higher when the source on the left is taken into account. As such, we have the small non-zero

---

[13] This serves as an extension of the demonstration in [20] of zero average transfer in a lattice of spatial and temporal period 2 using $k = 1$ to a domain of arbitrary period p.

[14] In some systems, it may be possible to establish a minimal k (related to the synchronisation time τ for the entropy rate [63]) to eliminate all information storage here. The situation is more complicated than [63] in being required to consider the source variable also.

$t(i, j = 1, n+1, k = 16) = 0.20$ bits to the right at this point in the periodic domain. This effect occurs for both the complete and apparent measures, as well as the temporal entropy rate, and we emphasise that it is not a finite-k effect. **These small non-zero transfer values are found regularly through the domain, and can be considered as an ambient transfer**. Note that they are stronger in the wake of a glider, indicating the absence of (relatively more common) following gliders. As a general rule though, where $t(i, j, n) < a(i, n)$ we can determine that the point is part of the background domain rather than a glider. On the other hand, if the transfer in the "wake" of a glider is stronger than the information storage at those points, then the "wake" should really be considered a glider structure itself.

Furthermore, there are larger information transfer values in the domain of rule 18 (see e.g. Fig. 3.9d), though these remain less significant than those in the domain walls there. There is an interesting pattern to these values, which we will discuss in regard to the interplay between the apparent and complete TE in Sect. 4.2.4.

Given these effects, we describe gliders as the *dominant*, as opposed to the only, information transfer entities here.

Finally, we note the implication that **there is no more information transfer in a system once it reaches a periodic or fixed-point attractor**. This is because the next state of each agent then becomes completely predictable from its past[15]: it is executing an information storage process. From the perspective of distributed computation, the system has completed an intrinsic computation of its final attractor and phase on it.

4.2.4 Apparent and Complete Transfer Entropy are Complementary

The apparent and complete transfer entropies have a similar nature and are complementary in together determining the next state of the destination (as in Eq. (4.56)). Importantly, they can reveal different aspects of the underlying dynamics. A simple example is shown on application to the rule ϕ_{par} (in Fig. 3.6), which has simple gliders moving at both 1 and 3 cells per time step. Here, the apparent TE does not appear to distinguish whether the primary information source of either glider type is the $j = 1$ or $j = 3$ channel. On the other hand, in conditioning out other sources the complete TE associates the slow moving gliders with the $j = 1$ channel and the fast moving gliders with the $j = 3$ channel.

---

[15] One could argue, following the preceding paragraphs in this section, that the system will still contain some ambient transfer while on the periodic attractor. The argument would be that in examining only a single cell's history (without knowing the attractor had been reached), one could not rule out the possibility that an unexpected glider could be encountered. We note however that this argument assumes *stationary statistics* in constructing the relevant PDFs using observations from both before and after the attractor is reached. We argue that without the prospect of escaping the attractor, the statistics should not be assumed to be stationary across this boundary. Thus, in computing stationary statistics taken from the attractor dynamics only, there are no more unexpected gliders to account for, and therefore there is no ambient transfer there.

This by no means implies that the apparent TE should be ignored in favour of the complete measure. In this section we discuss the more subtle differences revealed in measurements in directions orthogonal to particles, the information structure in the domain of rule 18, and in analysis of interaction-based transfer. These examples further reveal the complementary utility of both measures.

4.2.4.1 Information Transfer Orthogonal to Particles

When measured in the *orthogonal* direction to particles, the complete TE returns positive values at certain locations; e.g. for the γ^+ glider in the $t^c(i, j = -1, n + 1, k = 16)$ profile of rule 54 in Fig. 3.4g. In more detail, in between the points marked "x" in Fig. 3.8 there is non-zero t^c to the left although the γ^+ glider is actually moving to the right. This occurs for a similar reason to the non-zero transfer in the background domain, simply: the source does add information about the next state of the destination in addition to the previously conditioned sources. In Fig. 3.8 the source cell on the right (as part of the domain) informs an observer that there is no left-moving glider or static blinker to intercept the known right-moving glider. Importantly, this orthogonal transfer is not as significant as that in the macroscopic glider direction in terms of magnitude and coherence.

The apparent TE on the other hand can measure *negative* as well as positive values in particles moving orthogonal to the direction of measurement. For example, see the γ^- gliders in the $t(i, j = 1, n, k = 16)$ profile of rule 54 in Fig. 3.4e. As suggested in Sect. 4.1.3 such negative values occur when the probability of observing the actual next state of the destination is *lower* when the source on the left is taken into account than when it is not. Here the source, as part of the domain, suggests that this same domain found in the past of the destination is likely to continue; however since the next state of the destination is actually part of a glider, this suggestion proves to be *misinformative*. For example, consider the "x" marks in Fig. 3.8 which denote parts of the right-moving γ^+ gliders. For the source on the right (still in the domain), we have $p(x_{i,n+1} \mid x_{i,n}^{(k=16)}, x_{i+1,n}) = 0.13$ and $p(x_{i,n+1} \mid x_{i,n}^{(k=16)}) = 0.25$ giving $t(i, j = -1, n, k = 16) = -0.90$ bits: this is negative because the source (still in the domain) was misinformative about the next state of the destination. The complete TE was not negative for this same source (see the above paragraph) since it took into account the other causal source (on the left) that drove the glider. These negative values for the apparent measure are in fact useful, since they clearly identify orthogonal gliders while the complete measure does not.[16] Importantly again, the local apparent TE in the direction of glider motion was more informative than that in the orthogonal direction was misleading.

[16] Further utility for this misinformation is described in Chap. 5.

4.2.4.2 Information Structure and Interaction–Based Transfer in Domain of Rule 18

While both the apparent and complete transfer entropies measure strong information transfer in the domain walls of rule 18, they give different but complementary results in the background domain there (see Fig. 3.9).

As described in Sect. 3.3.5, this domain is of spatial and temporal period two, with every second site being "0" and every other site being either a "0" or a "1". Since the "0"'s at every second site are completely predictable (in the absence of domain walls) given their past history, h_μ at these points approaches zero bits. There is no diversity in state transitions of the destination here (following the interpretation in [37, 40]), so we have t and t^c approaching zero also.

On the other hand, at every other site $h_\mu(i, n+1, k)$ approaches 1 bit since observations of a "0" or a "1" are roughly equally likely with the past history indicating this alternate phase of the background. There is much scope or capacity for the sources to add information here, however the local apparent TE approaches zero bits at these sites. This is because the alternate sites are determined by an XOR interaction between both transfer sources [64]: as discussed in Sect. 4.1.3 prediction of this operation requires knowledge of *both* sources. Following [37, 40], although we have a high diversity of state transitions in the destination here, the high *assortative noise* between the source and destination obscures its contribution. As described in Sect. 4.1.3, **the apparent TE profiles vanish because there is no *coherent* transfer from the source to the destination in these strong interactions**. On the other hand, in conditioning on the opposite source **the complete transfer entropies are able to measure approximately 1 bit in predicting the outcome of these XOR operations**. As such, the t^c profiles are non-zero at only the alternate sites **because** (as described in Sect. 4.1.3) **this measure also detects interaction-based transfer**. Indeed, this example brings to mind discussion on the nature of information transfer in complex versus chaotic dynamics [5, 15–17]. It suggests that perhaps in chaotic dynamics, where many sources influence outcomes in an incoherent manner, the complete measure may indicate large information transfer whereas the apparent measure does not (because the interactions cannibalise coherent transfer effects).[17]

Finally, we note that it is not an anomaly that $t^c(i, j, n+1, k)$ for both $j = 1$ and -1 approach $h_\mu(i, n+1, k)$ at the alternate site. This is because the total information to predict any one site is summed from $t^c(i, j = 1, n+1, k)$ and $t(i, j = -1, n+1, k)$ or vice-versa (as per Eqs. (4.56) and (4.57)) not by summing the two complete transfer entropies from both sources. This example is a good demonstration of the manner in which the apparent and complete transfer entropies are complementary and both valuable, and how they fit with information storage in computing the next state of a variable.

[17] This suggestion is further investigated in an analysis of information transfer through an order-chaos phase transition in Chap. 6.

4.3 Information Flow as Causal Effect

That correlation is not causation is well-understood. Yet while authors increasingly consider the notions of information transfer and information flow and how they fit with our understanding of correlation and causality [20, 26, 29, 32–34, 53, 65], several questions nag. Is information transfer akin to causal effect? If not, what is the distinction between them? When examining the "effect" of one variable on another (e.g. between brain regions), should one seek to measure information transfer or causal effect? Despite the interest in this area, it remains unclear how the notion of information transfer should sit with the concept of causal effect.

As we have previously discussed in this chapter, predictive information transfer refers to the amount of information that a source variable adds to the next state of a destination variable; i.e. "if I know the state of the source, how much does that help to predict the state of the destination?".

Causal effect refers to the extent to which the source variable has a direct influence or drive on the next state of a destination variable, i.e. "if I change the state of the source, to what extent does that alter the state of the destination?". Information from causal effect can be seen to *flow* through the system, like injecting dye into a river [26]. In an Aristotelian sense, we restrict our interpretation to *efficient* cause here (e.g. see [66]).

Unfortunately, these concepts have become somewhat tangled in discussions of information transfer. Measures for both predictive transfer [20] and causal effect [26] have been inferred to capture information transfer in general, and measures of predictive transfer have been used to infer causality [27–30] with the two sometimes (problematically) directly equated (e.g. [31–36]).

The presentation of the information transfer component of distributed computation cannot be considered complete while such uncertainty clouds the concept. Our argument here is that the concepts of predictive transfer and causal effect are quite distinct: we aim to clarify their relationship and describe the manner in which they should be considered separately.

It is well-recognised that measurement of causal effect necessitates some type of *perturbation* or *intervention* of the source so as to detect the effect of the intervention on the destination (e.g. see [67, 68]). Attempting to infer causality without doing so leaves one measuring correlations of observations, regardless of how directional they may be [26]. In this section, we adopt the measure information flow for this purpose, and introduce a method for applying it on a local scale.

4.3.1 Information Flow

Following Pearl's probabilistic formulation of causal Bayesian networks [67], Ay and Polani [26] consider how to measure causal information flow via *interventional conditional probability distribution functions*. For instance, an interventional conditional PDF $p(a \mid \hat{s})$ considers the distribution of a resulting from *imposing* the value of $\hat{s}$. *Imposing* means intervening in the system to *set* the value of the imposed variable, and is at the essence of the definition of causal information flow. As an illustration of the difference between interventional and standard conditional PDFs, consider two correlated variables s and a: their correlation alters $p(a \mid s)$ in general from $p(a)$. If both variables are solely caused by another variable g however, then even where they remain correlated we have $p(a \mid \hat{s}) = p(a)$ because imposing a value $\hat{s}$ has no effect on the value of a.

In a similar fashion to the definition of transfer entropy as the deviation of a destination from *stochastic* independence on the source in the content of the destination's past, Ay and Polani propose the measure **information flow** as the deviation of the destination B from *causal* independence on the source A *imposing* another set of nodes $\mathbf{S}$. Mathematically, this is written as:

$$I_p(A \rightarrow B \mid \hat{\mathbf{S}}) = \sum_{\mathbf{s}} p(\mathbf{s}) \sum_{a} p(a \mid \hat{\mathbf{s}}) \sum_{b} p(b \mid \hat{a}, \hat{\mathbf{s}}) \log_2 \frac{p(b \mid \hat{a}, \hat{\mathbf{s}})}{\sum_{a'} p(a' \mid \hat{\mathbf{s}}) p(b \mid \hat{a}', \hat{\mathbf{s}})}. \tag{4.59}$$

The value of the measure is dependent on the choice of the set of nodes $\mathbf{S}$. It is possible to obtain a measure of apparent causal information flow $I_p(A \rightarrow B)$ from A to B without any $\mathbf{S}$ (i.e. $\mathbf{S} = \oslash$), yet this can be misleading. In particular, it ignores causal information flow arising from interactions of the source with another source variable. For example, if $b = a$ XOR s and $p(a, s) = 0.25$ for each combination of binary a and s, then $I_p(A \rightarrow B) = 0$ despite the clear causal effect of A, while $I_p(A \rightarrow B \mid \hat{S}) = 1$ bit. Also, we may have $I_p(A \rightarrow B) > 0$ only because A effects $\mathbf{S}$ which in turn effects B; where we are interested in *direct* causal information flow from A to B only $I_p(A \rightarrow B \mid \hat{\mathbf{S}})$ validly infers no direct causal effect.

Here we are interested in measuring the *direct* causal information flow from A to B, so we must either include all possible other sources in $\mathbf{S}$ or at least include enough sources to "block"[18] all non-immediate directed paths from A to B [26]. The minimum to satisfy this is the set of all direct causal sources of B excluding A, including any past states of B that are direct causal sources. For computing direct information flow across one cell to the right in ECAs where $a = x_{i-1,n}$ and $b = x_{i,n+1}$, this means $\mathbf{S}$ includes the immediate past of the destination cell and the previous state of the cell on its right $\left(\text{i.e. } \mathbf{s}_{i,n} = \{x_{i,n}, x_{i+1,n}\}\right)$. This is displayed in Fig. 4.5 (note the contrast to the transfer entropy in Fig. 4.2). Generalised as $I_p(j)$ for information flow across j cells to the right in any 1D CA, we have:

[18] A set of nodes U *blocks* a path of causal links where there is a node v on the path such that either:

1. $v \in U$ and the causal links through v on the path are not both into v, or
2. the causal links through v on the path are both into v, and v and all its causal descendants are not in U.

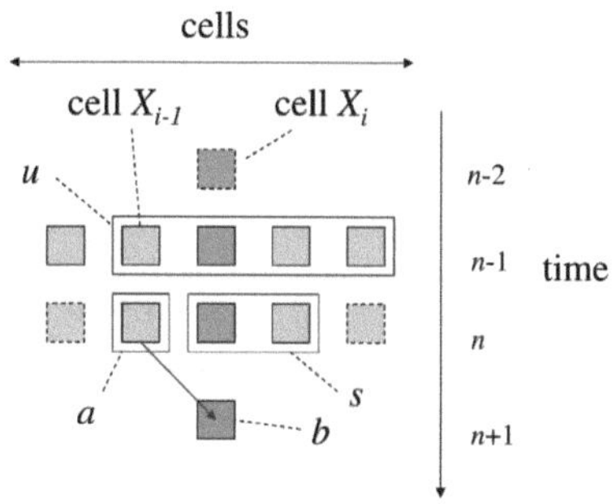

Fig. 4.5 Information flow across one cell to the right in ECAs $f(i, j = 1, n + 1)$: the contribution of a causal effect from source cell X_{i-1} to the next state of the destination cell X_i at time $n + 1$, imposing the previous states of the destination cell and the other information contributing cell X_{i+1}. We have the source $a = x_{i-1,n}$, the destination $b = x_{i,n+1}$, the imposed contributors are $\mathbf{s} = \{x_{i,n}, x_{i+1,n}\}$ and the cells blocking a back-door path (see Appendix D) relative to $(\mathbf{s}, a)$ are $u = \{x_{i-1,n-1}, x_{i,n-1}, x_{i+1,n-1}, x_{i+2,n-1}\}$. (NB: This figure was first published in [25]; reprinted with kind permission of the European Physical Journal (EPJ))

$$\mathbf{s}^j_{i,r,n} = \left\{x_{i,n}, \mathbf{v}^j_{i,r,n}\right\}. \tag{4.60}$$

The major task in computing $I_p(A \rightarrow B \mid \hat{\mathbf{S}})$ is the determination of the underlying interventional conditional PDFs in Eq. (4.59). By definition these may be gleaned by observing the results of intervening in the system, however this is not possible in many cases.

One alternative is to use detailed knowledge of the dynamics, in particular the structure of the causal links and possibly the underlying rules of the causal interactions. This also is often not available in many cases, and indeed is often the very goal for which one turned to such analysis in the first place. Regardless, where such knowledge is available it may allow one to make direct inferences. An important example is where the observed variable is known to be completely determined by the imposing set (e.g. $p(b \mid \hat{a}, \hat{\mathbf{s}})$ in ECAs in Fig. 4.5 can be determined as 0 or 1 from the CA rule table). Indeed, with $\mathbf{S}$ selected to compute *direct* information flow, B is determined from A and $\mathbf{S}$ (save for any underlying stochasticity), and one can use observational probabilities alone for $p(b \mid \hat{a}, \hat{\mathbf{s}})$ when all $\{a, \mathbf{s}\}$ combinations are observed. Another example is where the observed variable remains unaffected by the imposition (e.g. $p(a \mid \hat{\mathbf{s}})$ becomes $p(a)$ in ECAs in Fig. 4.5) allowing one to use the observational probabilities alone independently of the imposed variable.

Under certain constrained circumstances, one can construct these values from observational probabilities only [26], e.g. with the "back-door adjustment" (see Appendix D). A particularly important constraint on using the back-door adjustment here is that all $\{\mathbf{s}, a\}$ combinations must be observed.

4.3.2 Local Information Flow

We introduce a definition for a **local information flow**:

$$f(a \rightarrow b \mid \hat{\mathbf{s}}) = \log_2 \frac{p(b \mid \hat{a}, \hat{\mathbf{s}})}{\sum_{a'} p(a' \mid \hat{\mathbf{s}}) p(b \mid \hat{a}', \hat{\mathbf{s}})}, \tag{4.61}$$

in a similar manner to the other localisations performed following Sect. 2.2.2. The meaning of the local information flow is slightly different however. Certainly, it is an *attribution* of local causal effect of a on b were $\hat{\mathbf{s}}$ imposed at the given observation $(a, b, \mathbf{s})$. However, one must be aware that $I_p(A \rightarrow B \mid \hat{\mathbf{S}})$ is not the *average* of the local values $f(a \rightarrow b \mid \hat{\mathbf{s}})$ in exactly the same manner as the local values derived following Sect. 2.2.2. Unlike standard information-theoretical measures, the information flow is averaged over a product of *interventional* conditional probabilities $(p(\mathbf{s})p(a \mid \hat{\mathbf{s}})p(b \mid \hat{a}, \hat{\mathbf{s}})$, see Eq. (4.59)) which in general does not reduce down to the probability of the given observation $p(\mathbf{s}, a, b) = p(\mathbf{s})p(a \mid \mathbf{s})p(b \mid a, \mathbf{s})$. For instance, it is possible that not all of the tuples $\{a, b, \mathbf{s}\}$ will actually be observed, so averaging over observations would ignore the important contribution that any unobserved tuples provide to the determination of information flow. Again, the local information flow is specifically tied not to the *given observation* at time step n but to the *general configuration* $(a, b, \mathbf{s})$, and only *attributed* to the associated observation of this configuration at time n.

For lattice systems, we use the notation $f(i, j, n+1)$ to denote the local information flow into variable X_i from the source X_{i-j} at time step $n + 1$ (i.e. flow across j cells to the right). For CAs we have:

$$f(i, j, n+1) = \log_2 \frac{p(x_{i,n+1} \mid \widehat{x_{i-j,n}}, \widehat{\mathbf{s}^{j}_{i,r,n}})}{\sum_{x'_{i-j,n}} p(x'_{i-j,n} \mid \widehat{\mathbf{s}^{j}_{i,r,n}}) p(x_{i,n+1} \mid \widehat{x_{i-j,n}'}, \widehat{\mathbf{s}^{j}_{i,r,n}})}, \tag{4.62}$$

with $\mathbf{s}^{j}_{i,r,n}$ defined in Eq. (4.60).

4.4 Local Causal Information Flow in Cellular Automata

In this section we examine results from applying the local information flow to CAs in order to contrast the concepts of causal effect and information transfer. We focus on the results for rule 54 in Fig. 4.6, which displays the local information flow profile (Fig. 4.6b) of rule 54, as well as several transfer entropy profiles for comparison. We focus on transfer and flow one step to the right per unit time step $j = 1$, and measure the average transfer values being $T(j = 1, k = 16) = 0.080$ and $T^c(j = 1, k = 16) = 0.193$ bits for apparent and complete TE respectively, and the information flow at $I_p(j = 1) = 0.523$ bits. Importantly, much more insight is provided by

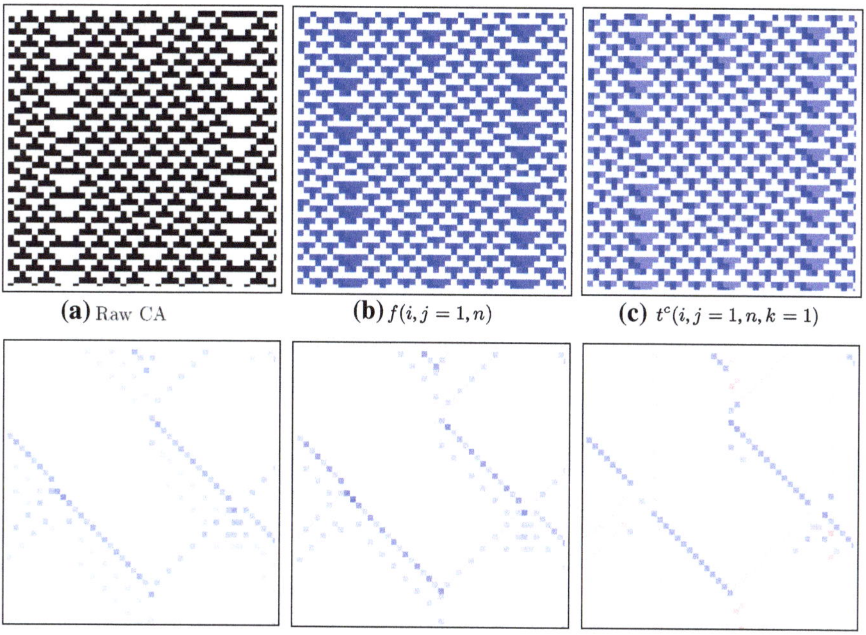

(a) Raw CA (b) $f(i, j = 1, n)$ (c) $t^c(i, j = 1, n, k = 1)$

(d) $t^c(i, j = 1, n, k = 16)$ (e) $t(i, j = 1, n, k = 16)$ (f) $t(i, j = 2, n, k = 16)$

Fig. 4.6 Local transfer entropy and information flow for raw states of rule 54 in **a** (45 time steps displayed for 45 cells, time increases down the page): **b** *Local information flow* across one cell to the right, max. 1.07 bits, min 0.00 bits; *Local complete transfer entropy* across one cell to the right: **c** with past history length $k = 1$, max. 1.17 bits, min 0.00 bits, and **d** past history length $k = 16$, max. 9.22 bits, min 0.00 bits; *Local apparent transfer entropy*: **e** across one cell to the right, max. 7.93 bits, min −4.04 bits, and **f** across two cells to the right, max. 6.00 bits, min 8.73 bits. (NB: This figure was first published in [25]; reprinted with kind permission of the European Physical Journal(EPJ))

contrasting the *local* values of each measure however, and we describe several cases within these results to highlight the differences in the concepts of information transfer and causal effect. These differences have been observed for transfer and flow one step to the left per unit time step (i.e. $j = -1$) also, and in other CAs with emergent structure (e.g. rules 110 and 18), and we comment on the generality of these results to other systems.

We demonstrate that causal effect is a fundamental system property, which is pervasive at the level of the micro-dynamics. We show in Sect. 4.4.1 that TE does not detect all of the causal effects that information flow does, while information flow does not detect the emergent computational structure (i.e. particles) that TE does. We show in Sect. 4.4.2 that transfer entropy can be interpreted as information transfer only when applied to directly causal information sources. We conclude that the information flow should be used in the first instance to study the causal structure, then the transfer entropy used to study emergent information transfer structure in distributed computation. Finally, in Sect. 4.4.3 we explore the conditions under which the transfer entropy converges with the concept of causal effect.

4.4.1 Information Transfer, Causal Flow and Emergent Structures

The most important result here is that the local information flow measures similar levels of causal effect in both the gliders and the background domain (see Fig. 4.6b in comparison to Fig. 4.6d, e). This is in contrast to the transfer entropy which, as we have seen, measures much stronger information transfer in the gliders as compared to the domain. *Both measures are correct, but from different perspectives* in measuring different concepts. There are two key general lessons here.

1. **Transfer entropy does not detect all causal effects that information flow does.** The four time step period of the (longest) sequences in the domain is longer than any one binary-state cell could produce alone—the cells rely on interaction with their neighbours to produce these *coupled periodic processes*. To achieve the long periods here, some information is stored in neighbours and retrieved after a few time steps (as described in Sect. 3.3.2). This is *necessarily* underpinned by the coupled causal effect between the neighbours. From another perspective, much of the background domain is highly causal simply because had one imposed values on the sources there the destinations would have changed; hence we find the strong patterns of information flow here. The concept of information transfer is focused on distributed computation and is not intended to capture causal effect where that causal effect underpins information storage processes instead.
2. In this manner, **information flow does not detect emergent computational structure that transfer entropy does** (i.e. particles in CAs). As described in Sect. 4.2.1 gliders are information transfer entities, because the cell states in the glider region provide much stronger predictive information about the next states in the direction of glider motion than do the previous states of the destination cells. In addition, we see the information transfer terms combining with information storage in computing the next state of the cell in e.g. Eq. (4.56). For these reasons, we say that predictive transfer is the concept that more closely aligned with the popularly understood concept of information transfer. From a causal perspective, the same CA rules or templates $\{a, \mathbf{s}\}$ executed in the glider are also executed elsewhere in the domain of the CA—while imposing the source value does indeed have a causal effect on the destination in the gliders, the positive directional information flow here is no greater than levels observed in the domain. The measure certainly captures the causal mechanism in the gliders, but its localisation does not distinguish that from the flow in the domain. In this form, the causal perspective focuses on the details or micro-level of the dynamics, whereas the predictive or computational perspective takes a

macroscopic view of emergent structures. It is possible that an explicitly macroscopic formulation of the information flow might distinguish gliders as highly causal macroscopic structures,[19] but certainly (when applied to the same source and destination pair as transfer entropy) as a directional measure of direct local causal effect it does not distinguish these emergent structures. On the other hand, the examination in the context of the past k states affords a macroscopic view to the TE, and emergent structure can only be detected on this scale. Since gliders are dislocations in background patterns [59] (in the past k states) which can only be caused by neighbouring cells, the source of the glider will add information about the destination in the context of this background pattern in the past k states (i.e. $p(x_{i,n+1} \mid x_{i,n}^{(k)}, x_{i-j,n}) > p(x_{i,n+1} \mid x_{i,n}^{(k)})$) and we have strong information transfer. On the other hand, information flow intrinsically cannot consider the context of the background patterns in the past, since imposing on $x_{i-j,n}$ and $\mathbf{s}_{i,r,n}^{j}$ blocks out the influence of those past k states.

4.4.2 *Information Transfer to be Measured from Causal Sources Only*

Figure 4.6f measures the local apparent TE $t(i, j = 2, n, k = 16)$ for two steps to the right per unit time step. This profile is fairly similar to that produced for one step to the right per unit time step (Fig. 4.6e). However, this measurement suggests a *superluminal* transfer, i.e. transfer from outside of the past light-cone of $x_{i,n}$ (see Sect. 2.2.4.1). The result is not intuitive as we expect zero information transfer from sources that are not direct causal information contributors. This is because only causal sources are present in Eq. (4.47) in contributing or *transferring* information to the next state of the destination. What we see in this profile merely reflects a correlation between the purported source and an actual causal source one cell away from the destination: the transfer entropy will produce a non-zero result from non-causal sources whenever such correlations exist. This does not mean that the TE measure is wrong, merely that it has not been correctly interpreted here. The key general result is that **in order to be genuinely interpreted as information transfer, the transfer entropy should only be applied to causal information sources for the given destination.** There it is a correlation which is *realised* in the computation that determines the next state of the destination. Beyond these sources though, it only measures correlations that do not directly contribute or transfer information into this computation.

To check the correctness of the information flow measure, we apply it here assuming the CA is of neighbourhood-5 (i.e. two causal contributors on either side of the destination with $r = 2$). As expected, the local information flow profile computes no causal effect across two cells to the right per unit time step (not shown). Importantly

[19] For example, perhaps looking at flow from a *set* of cells, in alignment with the perturbation-based measure the *local sensitivity* in [58].

however, note that the information flow could not be measured using observational data alone for either $j = 1$ or $j = 2$ in neighbourhood-5[20]; specific knowledge about the dynamics was required for the calculation.

Furthermore, measuring the complete TE $t^c(i, j = 2, n, k = 16)$ in this neighbourhood results in a zero predictive information profile (not shown). This is because all the information $h(i, n+1)$ required to predict the next state of the destination is contained within the interior $r = 1$ neighbourhood for this deterministic system. This information is represented in the right-hand side of Eq. (4.56), and is a subset of the conditioned variables in the expansion of $t^c(i, j = 2, n, k = 16)$. This measurement aligns well with the zero result for information flow. Significantly, only the complete TE could be applied for $j = 2$ using the available observational data alone, though both measures require the correct neighbourhood of other causal contributors $\mathbf{V}_X^Y$ to be a subset of those conditioned on or imposed here.

4.4.3 *Complete Transfer Entropy as an Inferrer for Information Flow*

The parallels between the complete TE and the information flow go beyond similar inference of a lack of influence. Consider the profile of $t^c(i, j = 1, n, k = 1)$ in Fig. 4.6c—note how similar it is to the profile of the local information flow in Fig. 4.6b. Indeed the average value $T^c(j = 1, k = 1) = 0.521$ bits is almost identical to the information flow $I_p(j = 1) = 0.523$ bits.

Convergence of the complete transfer entropy and direct information flow occurs with the combination of *one* **parameter setting and** *two* **conditions which are approximated in this example**:

1. the parameter k for the complete TE was set to include only the past states of the destination that are causal information contributors to its next state;
2. the condition that all $\{a, \mathbf{s}\}$ combinations are observed (this condition is relevant for averages but not local values); and
3. the condition that $p(a \mid \hat{\mathbf{s}}) \equiv p(a \mid \mathbf{s})$ (which for example is met where a is both causally and conditionally independent of $\mathbf{s}$).

We describe why these conditions lead to convergence in the following paragraphs.

With history length $k = 1$ here the numerators of the local measures Eq. (4.35) and Eq. (4.62) in fact become equal. This is enabled because with k set to include only the past states of the destination that are causal information contributors to its next state[21]—*no more, no less*—the complete TE *conditions* on the same variables that the direct information flow *imposes* upon. That is, as shown in Eq. (4.60) $\mathbf{s}_{i,r,n}^j$ in

[20] The CA here does not produce all of the required $\{s, a\}$ combinations for computing the required interventional probabilities with the back-door adjustment—see Appendix D for details.

[21] That is, with $k = 1$ in the CAs here, though for example in [29] where the elements in Henon maps are causally effected by their previous two states, $k = 2$ would be appropriate rather than the use of $k = 1$ there.

Eq. (4.62) refers to the same variables as $\{x_{i,n}^{(k)}, \mathbf{v}_{i,r,n}^{j}\}$ in Eq. (4.35) with k set in this manner. Importantly, this argument is valid for the non-lattice forms in Eq. (4.26) and Eq. (4.61) (i.e. $\mathbf{s}$ refers to the same variables as $\{x_n^{(k)}, \mathbf{v}_{x,n}^{y}\}$). Building on this enabling then, since we are measuring *direct* causal effect $\mathbf{S}$ includes all direct causal sources of B excluding A and so the interventional probability $p(b \mid \hat{a}, \hat{\mathbf{s}})$ is equivalent to the conditional probability $p(b \mid a, \mathbf{s})$ when the $\{a, \mathbf{s}\}$ combination is observed (as stated in Sect. 4.3.1). This parameter setting then ensures the numerators of the *local* measures Eqs. (4.35) and (4.62) are the same. **The history length parameter k therefore has an important role in moving the (complete) transfer entropy between measuring information transfer (at large k) and approximating causal effect (at minimal k).**

Note that for the combinations of $\{a, \mathbf{s}\}$ which are not observed, $p(b \mid a, \mathbf{s})$ is technically undefined. This is not relevant for local values of either measure (since the given $\{a, \mathbf{s}\}$ must have been observed), or the average complete TE, but for the information flow these terms revert to $p(b \mid \hat{a}, \hat{\mathbf{s}})$ and contribute additionally to the average. For convergence of the *averages* $T_{Y \to X}^{c}(k)$ and $I_p(A \to B \mid \hat{\mathbf{S}})$ only, it is thus required that all $\{a, \mathbf{s}\}$ combinations are observed. The condition is met in this example.

Consider now that if the condition $p(a \mid \hat{\mathbf{s}}) \equiv p(a \mid \mathbf{s})$ is also met, then the denominator of Eq. (4.61) becomes $p(b \mid \mathbf{s})$. With $\mathbf{s}$ referring to the same variables as $\{x_n^{(k)}, \mathbf{v}_{x,n}^{y}\}$, the denominator of Eq. (4.61) then matches Eq. (4.26), and in conjunction with the above conditions we have equality between the two local values and their averages. Importantly, this condition does not require all values of $\mathbf{s}$ to be observed for convergence of the averages, since $p(\mathbf{s})$ in Eq. (4.59) eliminates the contribution of any unobserved values of $\mathbf{s}$.

This final condition is approximated but not quite exactly met in the CA example. As described in Sect. 4.3.1, we have $p(a \mid \hat{\mathbf{s}}) \equiv p(a)$ here. This final condition would still be met if in fact $p(a) = p(a \mid \mathbf{s})$ (i.e. $p(y_n \mid x_n^{(k)}, \mathbf{v}_{x,n}^{y}) = p(y_n)$ in the notation for Eq. (4.26)). That is, there is a class of systems which satisfy this condition because the source is *both* causally and conditionally independent of the other causal contributors to the destination. The CA example approximates the sub-condition $p(a \mid \mathbf{s}) = p(a)$. In Fig. 4.5 we see that while both $a = x_{i-1,n}$ and $\mathbf{s} = \left\{x_{i,n}, x_{i+1,n}\right\}$ have two common sources $\left(\left\{x_{i-1,n-1}, x_{i-1,n}\right\}\right)$, a has one extra and $\mathbf{s}$ has two extra sources that are not shared. It is these unshared sources that cannibalise the correlation between $\mathbf{s}$ and a. The small correlation here is confirmed by the Kullback-Leibler divergence (see [51]) of $p(a \mid \mathbf{s})$ from $p(a)$ (i.e. the mutual information between a and $\mathbf{s}$) which is very low (0.01 bits) for $j = \{1, -1\}$ for rule 54 here. The divergence is still low, but larger for other ECA rules with emergent structure (i.e. 0.03 bits for rule 110 and 0.13 bits for rule 18). Nonetheless, the non-zero divergence confirms that the condition is not precisely met. Finally we note that where the previous conditions (including $p(a \mid \hat{\mathbf{s}}) \equiv p(a)$) were met, the difference between the local values due to $p(a \mid \mathbf{s}) \neq p(a)$ may be written as:

$$\log_2 \frac{\sum_{a'} p(a') p(b \mid a', \mathbf{s})}{p(b \mid \mathbf{s})}, \tag{4.63}$$

or for the CA as:

$$\log_2 \frac{\sum_{x'_{i-j,n}} p(x'_{i-j,n}) p(x_{i,n+1} \mid x'_{i-j,n}, \mathbf{s}^j_{i,r,n})}{p(x_{i,n+1} \mid \mathbf{s}^j_{i,r,n})}. \tag{4.64}$$

Interestingly this difference is independent of the source value, $a = x_{i-j,n}$.

As such, where one cannot intervene in the system, and does not have the required observations to use a method such as the back-door adjustment (see Appendix D), the local complete TE could provide a useful inference for the local information flow profile. We explore this possibility in more detail in Appendix E. Despite the obvious utility of the complete TE here, we emphasise that it does measure a different concept and only infers the causal flow completely correctly when the above conditions are met.

4.5 Summary

In this section, we have described how the local transfer entropy quantifies the information transfer at space-time points within a system. Local transfer entropy presents insights that cannot be obtained using the averaged measure alone, in particular in providing these spatiotemporal information transfer profiles as an analytic tool. Most importantly though, they provide the first direct quantitative evidence that particles are the dominant information transfer entities in CAs. This result is important in bringing together the quantitative definition of information transfer (transfer entropy) with the popular understanding of the concept. It is also important because of analogies between particles in CAs and coherent structure or hypothesised information transfer entities in physical systems, such as travelling localisations caused by dipole-dipole interactions in microtubules [8] and in soliton dynamics [24].

The local view also allowed us to study the transfer entropy measure itself, including the importance of appropriate destination conditioning lengths k (e.g. that using $k \to \infty$ is most correct). It also allowed us to contrast the apparent and complete forms which were introduced here in order to explore how incorporating source interactions alters information transfer measurements (see definitions in Table 4.1). Furthermore, the local view revealed the manner in which information transfer fits with information storage in the distributed computation of the next state of a variable. We also demonstrated the important situation where the local apparent measure becomes negative at orthogonally moving particles (see a summary of information transfer properties of emergent structures in CAs in Table 4.2).

We have also demonstrated the complementary nature of the concepts of information transfer and causal information flow while emphasising the distinctions between

Table 4.1 Local measures relevant to information transfer

Measure	Local definition	Equation no.
Apparent transfer entropy	$t_{Y\to X}(n+1,k) = i(y_n; x_{n+1} \mid x_n^{(k)})$	Eq. (4.11)
Conditional transfer entropy	$t_{Y\to X\mid Z}(n+1,k) = i(y_n; x_{n+1} \mid x_n^{(k)}, z_n)$	Eq. (4.21)
Complete transfer entropy	$t^c_{Y\to X}(n+1,k) = i(y_n; x_{n+1} \mid x_n^{(k)}, \mathbf{v}^y_{x,n})$	Eq. (4.29)
Collective transfer entropy	$t_X(n+1,k) = i(\mathbf{v}_{x,n}; x_{n+1} \mid x_n^{(k)})$	Eq. (4.52)
Information flow	$f(a \to b \mid \hat{\mathbf{s}}) = \log_2 \frac{p(b\mid\hat{a},\hat{\mathbf{s}})}{\sum_{a'} p(a'\mid\hat{\mathbf{s}})p(b\mid\hat{a}',\hat{\mathbf{s}})}$	Eq. (4.61)

Table 4.2 Examples of emergent structures in cellular automata with specific information storage and transfer properties

	$t(i,j,n) < 0$	$t(i,j,n) > 0$
$a(i,n) > 0$		$a(i,n) > t(i,j,n)$: periodic domain with ambient transfer (Sect. 4.2.3) $t(i,j,n) > a(i,n)$: wake segment of particle (Sect. 4.2.3)
$a(i,n) < 0$	particle orthogonal to channel j (Sect. 4.2.4.1)	particles in channel j(Sect. 4.2.2)

them. Causal effect is a fundamental micro-level property of a system. Information flow should be used as a primary tool (where possible) to establish the presence of and quantify causal relationships. Information transfer can then be analysed in order to gain insight into the emergent computation being carried out by the system. Transfer entropy can be measured for any time-series pair, but can only be interpreted as a physical information transfer when measured on a direct causal link. We also showed the conditions under which the complete transfer entropy converges with the information flow. This included demonstrating the importance of k in tuning the transfer entropy between a measure of information transfer (at large k) and inferring causal effect (at small k). More generally, we noted that the use of a large k is critical to our perspective of distributed computation in general, in also separating information transfer from storage.

Finally, the local transfer entropy provides similar *filtering* for coherent structure in CAs to other methods [6, 58–62], yet is novel in a number of ways. The most important distinction with previous methods is that this filtering is part of a framework for the local information dynamics of distributed computation, combining the perspectives of information transfer with the information storage measures that we have already seen in Chap. 3, and the information modification we will see in Chap. 5. No other method uses multiple filtering perspectives; more importantly, none can decompose computation into its component operations. While all methods highlight particles along with other structure, this is the only filter that can provide quantitative evidence that particles are the dominant information transfer entities. Transfer entropy provides multiple filters itself here: one for each generic channel

j, and the complementary apparent and complete measures. These filters are distinct in focussing on the leading glider edges facilitating the information transfer and only the minimal part of domain walls necessary to identify them (from a temporal perspective). They are also distinct in filtering only moving structure, in particular information "moving" in the given channel. Also, the collective transfer entropy (equivalent to the local temporal entropy rate here) provides a useful single filter for all moving information. We comment further on the unique combination of filtering properties shared by all of the measures in this framework in our final summary in Sect. 9.1.5.

Together with the results from Chap. 3, we have now quantified both information storage and transfer in our framework for the local information dynamics of distributed computation. We have also described the manner in which these dynamics are embodied in CAs in blinkers, gliders and domain walls. However, we have not yet separately identified collision events in CAs. To complete our framework, we now consider the nature of information modification.

References

1. M. Mitchell, J.P. Crutchfield, P.T. Hraber, Evolving cellular automata to perform computations: Mechanisms and impediments. Physica D **75**, 361–391 (1994)
2. M. Mitchell, J.P. Crutchfield, R. Das, Evolving cellular automata with genetic algorithms: a review of recent work, in *Proceedings of the First International Conference on Evolutionary Computation and Its Applications, Moscow*, ed. by E.D. Goodman, W. Punch, V. Uskov (Russian Academy of Sciences, Russia, 1996)
3. M. Mitchell, Computation in cellular automata: a selected review, in *Non-Standard Computation*, ed. by T. Gramss, S. Bornholdt, M. Gross, M. Mitchell, T. Pellizzari (VCH Verlagsgesellschaft, Weinheim, 1998), p. 140
4. W. Hordijk, C.R. Shalizi, J.P. Crutchfield, Upper bound on the products of particle interactions in cellular automata. Physica D **154**(3–4), 240–258 (2001)
5. C.G. Langton, Computation at the edge of chaos: phase transitions and emergent computation. Physica D **42**(1–3), 12–37 (1990)
6. A. Wuensche, Classifying cellular automata automatically: finding gliders, filtering, and relating space-time patterns, attractor basins, and the Z parameter. Complexity **4**(3), 47–66 (1999)
7. S. Wolfram, Cellular automata as models of complexity. Nature **311**(5985), 419–424 (1984)
8. J.A. Brown, J.A. Tuszynski, A review of the ferroelectric model of microtubules. Ferroelectrics **220**, 141–156 (1999)
9. D.E. Edmundson, R.H. Enns, Fully 3-dimensional collisions of bistable light bullets. Opt. Lett. **18**, 1609–1611 (1993)
10. I. Sendiña-Nadal, E. Mihaliuk, J. Wang, V. Pérez-Muñuzuri, K. Showalter, Wave propagation in subexcitable media with periodically modulated excitability. Phys. Rev. Lett. **86**(8), 1646 (2001)
11. I. Couzin, R. James, D. Croft, J. Krause, Social organization and information transfer in schooling fishes, in *Fish Cognition and Behavior*, ser. Fish and Aquatic Resources, ed. by B.C.K. Laland, J. Krause (Blackwell Publishing, 2006), pp. 166–185
12. D.P. Varn, J.P. Crutchfield, From finite to infinite range order via annealing: the causal architecture of deformation faulting in annealed close-packed crystals. Phys. Lett. A **324**(4), 299–307 (2004)

13. A.S. Klyubin, D. Polani, C.L. Nehaniv, All else being equal be empowered, in *Proceedings of the 8th European Conference on Artificial Life (ECAL, 2005), Kent, UK*, ser. Lecture Notes in Computer Science, ed. by M.S. Capcarrere, A.A. Freitas, P.J. Bentley, C.G. Johnson, J. Timmis vol. 3630, (Springer, Berlin, 2005), pp. 744–753
14. M. Lungarella, O. Sporns, Mapping information flow in sensorimotor networks. PLoS Comput. Biol. **2**(10), e144 (2006)
15. R.V. Solé, S. Valverde, Information transfer and phase transitions in a model of internet traffic. Physica A **289**(3–4), 595–605 (2001)
16. O. Miramontes, Order-disorder transitions in the behavior of ant societies. Complexity **1**(3), 56–60 (1995)
17. D. Coffey, Self-organization, complexity and chaos: the new biology for medicine. Nat. Med. **4**(8), 882–885 (1998)
18. C.R. Shalizi, K.L. Shalizi, R. Haslinger, Quantifying self-organization with optimal predictors. Phys. Rev. Lett. **93**(11), 118701 (2004)
19. M.H. Jakubowski, K. Steiglitz, R. Squier, Information transfer between solitary waves in the saturable Schrödinger equation. Phys. Rev. E **56**(6), 7267 (1997)
20. T. Schreiber, Measuring information transfer. Phys. Rev. Lett. **85**(2), 461–464 (2000)
21. K. Lindgren, M.G. Nordahl, Complexity measures and cellular automata. Complex Syst. **2**(4), 409–440 (1988)
22. J.T. Lizier, M. Prokopenko, A.Y. Zomaya, Information transfer by particles in cellular automata, in *Proceedings of the Third Australian Conference on Artificial Life, Gold Coast, Australia*, ser. Lecture Notes in Artificial Intelligence, ed. by M. Randall, H.A. Abbass, J. Wiles, vol. 4828, (Springer, Berlin, 2007), pp. 49–60
23. J.T. Lizier, M. Prokopenko, A.Y. Zomaya, Local information transfer as a spatiotemporal filter for complex systems. Phys. Rev. E **77**(2), 026110 (2008)
24. J.K. Park, K. Steiglitz, W.P. Thurston, Soliton-like behavior in automata. Physica D **19**(3), 423–432 (1986)
25. J.T. Lizier, M. Prokopenko, Differentiating information transfer and causal effect. Eur. Phys. J. B **73**(4), 605–615 (2010)
26. N. Ay, D. Polani, Information flows in causal networks. Adv. Complex Syst. **11**(1), 17–41 (2008)
27. H. Sumioka, Y. Yoshikawa, M. Asada, Causality detected by transfer entropy leads acquisition of joint attention, in *Proceedings of the 6th IEEE International Conference on Development and Learning (ICDL 2007)*, (IEEE, London 2007), pp. 264–269
28. M. Vejmelka, M. Palus, Inferring the directionality of coupling with conditional mutual information. Phys. Rev. E **77**(2), 026214 (2008)
29. M. Lungarella, K. Ishiguro, Y. Kuniyoshi, N. Otsu, Methods for quantifying the causal structure of bivariate time series. Int. J. Bifurcat. Chaos **17**(3), 903–921 (2007)
30. P.F. Verdes, Assessing causality from multivariate time series. Phys. Rev. E **72**(2), 026 222–229 (2005)
31. T.Q. Tung, T. Ryu, K.H. Lee, D. Lee, Inferring gene regulatory networks from microarray time series data using transfer entropy, in *Proceedings of the Twentieth IEEE International Symposium on Computer-Based Medical Systems (CBMS '07), Maribor, Slovenia*, ed. by P. Kokol, V. Podgorelec, D. Mičetič-Turk, M. Zorman, M. Verlič (IEEE, Los Alamitos, USA, 2007), p. 388
32. K. Ishiguro, N. Otsu, M. Lungarella, Y. Kuniyoshi, Detecting direction of causal interactions between dynamically coupled signals. Phys. Rev. E **77**(2), 026216 (2008)
33. X.S. Liang, Information flow within stochastic dynamical systems. Phys. Rev. E **78**(3), 031113 (2008)
34. K. Hlaváčková-Schindler, M. Paluš, M. Vejmelka, J. Bhattacharya, Causality detection based on information-theoretic approaches in time series analysis. Phys. Rep. **441**(1), 1–46 (2007)
35. G. Van Dijck, J. Van Vaerenbergh, M.M. Van Hulle, Information theoretic derivations for causality detection: application to human gait, in *Proceedings of the International Conference on Artificial Neural Networks (ICANN 2007), Porto, Portugal*, ser. Lecture Notes in Computer

Science, ed. by J.M.D. Sá, L.A. Alexandre, W. Duch, D. Mandic, vol. 4669. (Springer, Berlin, 2007), pp. 159–168
36. Y.-C. Hung, C.-K. Hu, Chaotic communication via temporal transfer entropy. Phys. Rev. Lett. **101**(24), 244102 (2008)
37. R.V. Solé, S. Valverde, Information theory of complex networks: on evolution and architectural constraints, in *Complex Networks*, ser. Lecture Notes in Physics, ed. by E. Ben-Naim, H. Frauenfelder, Z. Toroczkai, vol. 650, (Springer, Berlin, 2004), pp. 189–207
38. C.W.J. Granger, Investigating causal relations by econometric models and cross-spectral methods. Econometrica **37**, 424–438 (1969)
39. A. Kaiser, T. Schreiber, Information transfer in continuous processes. Physica D **166**(1–2), 43–62 (2002)
40. M. Prokopenko, F. Boschietti, A.J. Ryan, An information-theoretic primer on complexity, self-organization, and emergence. Complexity **15**(1), 11–28 (2009)
41. F. Takens, Detecting strange attractors in turbulence, in *Dynamical Systems and Turbulence, Warwick 1980*, ser. Lecture Notes in Mathematics, ed. by D. Rand, L.-S. Young (Springer, Berlin, 1981), pp. 366–381
42. S.K. Baek, W.-S. Jung, O. Kwon, H.-T. Moon, Transfer entropy analysis of the stock market, 2005, arXiv:physics/0509014v2. http://arxiv.org/abs/physics/0509014
43. L.J. Moniz, E.G. Cooch, S.P. Ellner, J.D. Nichols, J.M. Nichols, Application of information theory methods to food web reconstruction. Ecol. Model. **208**(2–4), 145–158 (2007)
44. M. Chávez, J. Martinerie, M. Le Van Quyen, Statistical assessment of nonlinear causality: application to epileptic EEG signals. J. Neurosci. Methods **124**(2), 113–128 (2003)
45. J. Pahle, A.K. Green, C.J. Dixon, U. Kummer, Information transfer in signaling pathways: a study using coupled simulated and experimental data. BMC Bioinform. **9**, 139 (2008)
46. N. Bertschinger, E. Olbrich, N. Ay, J. Jost, Information and closure in systems theory, in *Proceedings of the 7th German Workshop on Artificial Life (GWAL-7), Jena, Germany*, ed. by S. Artmann, P. Dittrich (IOS Press, Amsterdam, 2006)
47. T. Helvik, K. Lindgren, M.G. Nordahl, Continuity of information transport in surjective cellular automata. Commun. Math. Phys. **272**(1), 53–74 (2007)
48. J.L. Massey, Causality, feedback and directed information, in *Proceedings of the International Symposium on Information Theory and its Applications* (Waikiki, Hawaii, USA, 1990)
49. P. Rämö, S. Kauffman, J. Kesseli, O. Yli-Harja, Measures for information propagation in Boolean networks. Physica D **227**(1), 100–104 (2007)
50. J.T. Lizier, M. Prokopenko, A.Y. Zomaya, The information dynamics of phase transitions in random Boolean networks, in *Proceedings of the Eleventh International Conference on the Simulation and Synthesis of Living Systems (ALife XI), Winchester, UK*, ed. by S. Bullock, J. Noble, R. Watson, M.A. Bedau (MIT Press, Cambridge, 2008), p. 381
51. D.J. MacKay, *Information Theory, Inference, and Learning Algorithms* (Cambridge University Press, Cambridge, 2003)
52. V.A. Vakorin, O.A. Krakovska, A.R. McIntosh, Confounding effects of indirect connections on causality estimation. J. Neurosci. Methods **184**(1), 152–160 (2009)
53. N. Lüdtke, S. Panzeri, M. Brown, D.S. Broomhead, J. Knowles, M.A. Montemurro, D.B. Kell, Information-theoretic sensitivity analysis: a general method for credit assignment in complex networks. J. Roy. Soc. Interface **5**(19), 223–235 (2008)
54. L.M.A. Bettencourt, V. Gintautas, M.I. Ham, Identification of functional information subgraphs in complex networks. Phys. Rev. Lett. **100**(23), 238701 (2008)
55. T.M. Cover, J.A. Thomas, *Elements of Information Theory* (Wiley, New York, 1991)
56. J.P. Crutchfield, D.P. Feldman, Regularities unseen, randomness observed: levels of entropy convergence. Chaos **13**(1), 25–54 (2003)
57. J.E. Hanson, J.P. Crutchfield, Computational mechanics of cellular automata: an example. Physica D **103**(1–4), 169–189 (1997)
58. C.R. Shalizi, R. Haslinger, J.-B. Rouquier, K.L. Klinkner, C. Moore, Automatic filters for the detection of coherent structure in spatiotemporal systems. Phys. Rev. E **73**(3), 036104 (2006)

59. J.E. Hanson, J.P. Crutchfield, The attractor-basin portait of a cellular automaton. J. Stat. Phys. **66**, 1415–1462 (1992)
60. T. Helvik, K. Lindgren, M.G. Nordahl, Local information in one-dimensional cellular automata, in *Proceedings of the International Conference on Cellular Automata for Research and Industry, Amsterdam*, ser. Lecture Notes in Computer Science, ed. by P.M. Sloot, B. Chopard, A.G. Hoekstra, vol. 3305, (Springer, Berlin, 2004), pp. 121–130
61. P. Grassberger, New mechanism for deterministic diffusion. Phys. Rev. A **28**(6), 3666 (1983)
62. P. Grassberger, Information content and predictability of lumped and distributed dynamical systems. Physica Scripta **40**(3), 346 (1989)
63. D.P. Feldman, J.P. Crutchfield, Synchronizing to periodicity: the transient information and synchronization time of periodic sequences. Adv. Complex Syst. **7**(3–4), 329–355 (2004)
64. P. Grassberger, Some more exact enumeration results for 1d cellular automata. J. Phys. A Math. General **20**(12), 4039–4046 (1987)
65. G. Auletta, G.F.R. Ellis, L. Jaeger, Top-down causation by information control: from a philosophical problem to a scientific research programme. J. Roy. Soc. Interface **5**(27), 1159–1172 (2008)
66. H.B. Veatch, *Aristotle: A contemporary appreciation* (Indiana University Press, Bloomington, 1974)
67. J. Pearl, *Causality: Models, Reasoning, and Inference* (Cambridge University Press, Cambridge, 2000)
68. L.R. Hope, K.B. Korb, An information-theoretic approach to causal power, Clayton School of Information Technology, Monash University, Technical Report 2005/176, 2005

Chapter 5
Information Modification

Information modification is often colloquially described as *processing*. It has been viewed as a particularly important operation for biological neural networks and models thereof [1–4], where it has been suggested as a potential biological driver [2]. It is also a key operation in collision-based computing (e.g. [5, 6], including soliton dynamics and collisions [7]).

Information modification has been interpreted to mean *interactions* between transmitted and/or stored information which result in a *modification* of one or the other [8]. We accept this interpretation in our perspective of distribution computation, as it specifically juxtaposes modification against storage and transfer, viewing it as a dynamic combination or synthesis of information from different sources. Modification therefore involves a *non-trivial* processing of information rather than a trivial movement or translation of one source of information.

The term has however remained elusive to appropriate quantitative definition, despite several attempts [1, 3, 4]. A major issue is that these attempts focus on measuring *trivial* processing as the movement or interpretation of information rather than specifically the *modification* of information synthesised from several sources. Furthermore, some [1, 4] have been too specific to allow portability across system types (e.g. by focusing on the capability of a system to solve a known problem, or measuring properties related to the particular type of system being examined), or are not amenable to measuring information modification at *local* space-time points *within* a distributed system.

To derive quantitative insights here, we again focus on CAs where (as per Sect. 2.3) the qualitative notion of information modification in distributed computation is well-understood. In this context, *interactions* between transmitted and/or stored information is generally interpreted to mean collisions of particles (including blinkers as information storage), with the resulting dynamics involving something other than the incoming particles continuing unperturbed. The resulting dynamics could involve zero or more particles (with an annihilation leaving only a background domain), perhaps including some of the incoming particles.[1] Given the focus on perturbations in

[1] The number of particles resulting from a collision has been studied elsewhere [9].

J. T. Lizier, *The Local Information Dynamics of Distributed Computation in Complex Systems*, Springer Theses, DOI: 10.1007/978-3-642-32952-4_5,
© Springer-Verlag Berlin Heidelberg 2013

the definition here, it is logical to associate a collision event with the modification of transmitted and/or stored information, and to see it as an information processing or decision event. Indeed, as an information processing event the important role of particle collisions in determining future dynamics is widely acknowledged for CAs [8–12], e.g. in the ϕ_{par} density classification task [13, 14], and is paralleled in studies of collision-based computing [5–7].

In this chapter, we aim to quantify information modification on a local scale in space and time, in order to complete our framework for the information dynamics of distributed computation. We hypothesise that such a measure should identify particle collisions in CAs as the dominant information modification events. We derive the separable information in Sect. 5.1 as a tool to detect *non-trivial information modification events*, where separate inspection of information sources is misinformative about the next state of a destination. Indeed, we demonstrate in Sect. 5.2 that the separable information is the first measure to identify collisions in CAs as such non-trivial information modification events.[2]

Additionally, we quantify *irreversible information destruction* on a local scale in distributed computation in Sect. 5.3 so as to explore the relationship of this concept to information modification. We apply this measure to CAs, where reversibility has often been considered yet information destruction has not previously been measured. Our local measurements here demonstrate that the concept of information destruction is complementary to but distinct from information modification and its associated events. Certainly many particle collisions involve information destruction, yet we observe modification to occur without destruction, and destruction to occur without modification.

5.1 Separable Information as a Detector for Non-Trivial Information Modification

We begin our investigation of the quantitative nature of information modification by considering what it means for a particle in a CA to be *modified.* For the simple case of a glider, a modification is simply an alteration to the predictable periodic pattern of the glider's dynamics. At such points, an observer would be *surprised* or *misinformed* about the next state of the perturbed glider, having not taken account of the entity about to perturb it.

This interpretation is a clear reminder of our earlier comments that local active information storage was misinformative at moving gliders (Sect. 3.3.3), and local apparent transfer entropy was misinformative at gliders travelling in the orthogonal direction to the measurement (Sect. 4.2.4.1). For these unperturbed gliders, one expects the local apparent transfer entropy measured in the direction of motion to be *more informative* about its continuation than any misinformation conveyed from other sources (see Fig. 5.1a). However, where the glider is modified by a collision

[2] The separable information and its application to CAs were first reported in [15, 16].

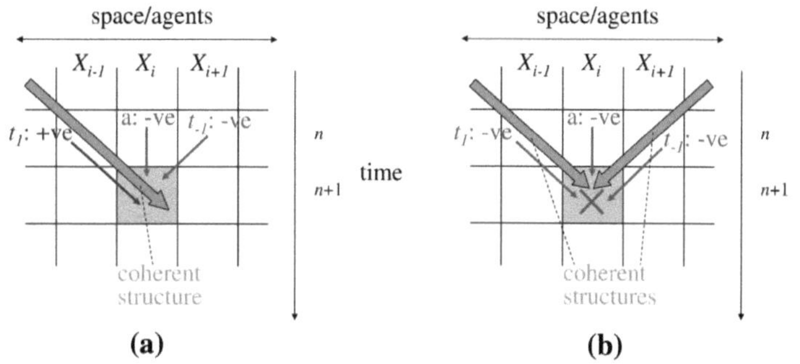

Fig. 5.1 Our expectations for the local information dynamics of storage and transfer for unperturbed gliders (coherent structures) and glider collisions. **a** For *unperturbed gliders* in channel $j = 1$, we expect the transfer $t(i, j = 1, n, k)$ in the direction of glider motion to be positively informative, and indeed more informative than the misinformation conveyed through a and $t(i, j = -1, n, k)$. **b** For a *collision perturbing a glider* moving in the channel $j = 1$, we can no longer expect the transfer $t(i, j = 1, n, k)$ in the direction of glider motion to be positively informative at the collision point. We cannot expect the transfer $t(i, j = -1, n, k)$ in the direction of the incident glider to be positively informative at the collision point either. (NB: Reprinted with permission from J. T. Lizier et al. [16] Copyright 2010, American Institute of Physics)

with another glider, we can no longer expect the local apparent transfer entropy in its macroscopic direction of motion to remain informative about its evolution (see Fig. 5.1b). Assuming that the incident glider is also perturbed, the local apparent transfer entropy in its macroscopic direction of motion will also not be informative about its evolution at this collision point. We expect the same argument to be true for domain walls and their collisions.

As such, we make the hypothesis that at the spatiotemporal location of a local information modification event or collision, *separate* inspection of each information source will *misinform* an observer overall about the next state of the modified information destination. Such separate inspection contrasts with the information gained from a unified inspection of all causal sources (i.e. $I_{\mathbf{V}_X;X}$), where the observer can account for the interaction of the sources producing a modification.

To be specific, the information sources referred to here are the past history of the destination (via the local active information storage from Chap. 3) and each other causal information contributor: these are examined in the context of the past history of the destination, via their local apparent transfer entropies from Chap. 4.

We have seen how these sources provide the total information for the computation of the next state of the destination when interactions are accounted for in Eq. (4.54.).[3] In contrast, we quantify the total information gained from *separate* observation of the information storage and information transfer contributors as the **local separable information** $s_X(n)$:

[3] Interactions are accounted in that equation because of the incremental conditioning on previous sources (i.e. in conditional and complete transfer entropies). Consideration of apparent transfer entropies only in Eq. (5.2)means that interactions are not being accounted for.

$$s_X(n) = \lim_{k\to\infty} s_X(n,k), \tag{5.1}$$

$$s_X(n,k) = a_X(n,k) + \sum_{Y\in \mathbf{V}_X \setminus X} t_{Y\to X}(n,k), \tag{5.2}$$

with the subscripts indicating the destination X and source variables $Y \in \mathbf{V}_X \setminus X$. We use $s_X(n,k)$ for finite-k estimates, though in practise recommend that as large a k as possible is used. This is because the measure relies on the correctness of $a_X(n,k)$ and the $t_{Y\to X}(n,k)$, which have this requirement as we have shown in Chaps. 3 and 4. Indeed, the true separation of elements of information storage and transfer (facilitated by $k \to \infty$ providing the context of the past) is critical to our consideration of information modification from the perspective of distributed computation here.[4]

As an example, in Fig. 4.1. we have $s_X(n,k) = a_X(n,k) + t_{Y_1\to X}(n,k) + t_{Y_2\to X}(n,k)$. For CAs, where the causal information contributors are homogeneously within the neighbourhood r, we write the local separable information in lattice notation as:

$$s(i,n) = \lim_{k\to\infty} s(i,n,k), \tag{5.3}$$

$$s(i,n,k) = a(i,n,k) + \sum_{j=-r, j\neq 0}^{+r} t(i,j,n,k). \tag{5.4}$$

We show $s(i,n,k)$ diagrammatically in Fig. 5.2.

As inferred in our hypothesis, we expect the local separable information to be *positive* or *highly separable* where separate observations of the information contributors are informative overall regarding the next state of the destination. This may be interpreted as a *trivial* information modification, because an observer is positively informed even without accounting for any interactions between the sources. As such, information storage and transfer are not interacting in any significant manner. For example, a periodic process executes information storage alone, trivially updating its state using its past history as a sole information source.

More importantly, we expect the local separable information to be *negative* or *non-separable* at spatiotemporal points where an information modification event or collision takes place. Here, separate observations are misleading overall because the outcome is largely determined by the interaction of the information sources. We say that a **non-trivial information modification** is taking place, understanding this as the interaction between information storage and transfer. For example, a particle collision involves a non-trivial information modification because an observer needs

[4] The separable information has parallels to the sum of "first order terms" (the apparent contribution of each source without considering interactions) in the total information of a destination in [17]. The local separable information however is distinguished in considering the contributions in the context of the destination's past, and evaluating these on a local scale—both features are critical for an understanding of distributed computation.

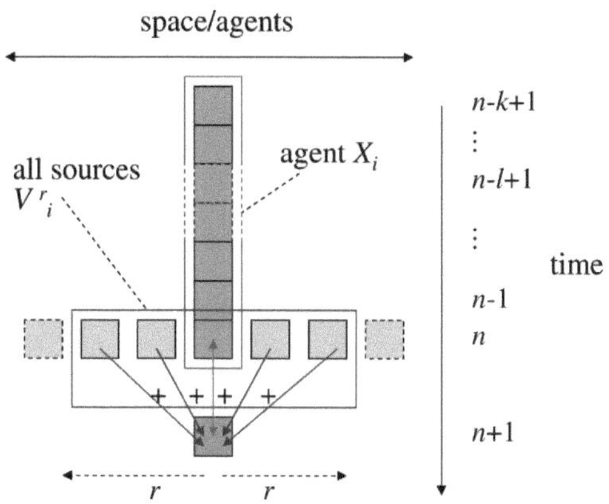

Fig. 5.2 Separable information $s(i, n + 1, k)$: information gained about the next state of the destination $x_{i,n+1}$ from separately examining each causal information source in the context of the destination's past $x^{(k)}_{i,n}$. For ECAs these causal sources are within the cell range r. (NB: Reprinted with permission from J. T. Lizier et al. [16]. Copyright 2010, American Institute of Physics)

to collectively examine multiple sources and their interaction in order to be positively informed about the next state of the destination.

Interestingly, this formulation of non-trivial information modification echoes descriptions of complex systems as consisting of (a large number of) elements interacting in a *non-trivial* fashion [18], and of emergence as where *"the whole is greater than the sum of its parts"* [19]. Here, we quantify the sum of the parts in $s(i, n)$, whereas "the whole" refers to examining all information sources together. The whole is greater where all information sources must be examined together in order to receive positive information on the next state of the examined entity. We emphasise, there is no quantity representing "the whole" as such, simply the indication that the sources must be examined *together*. We also emphasise that $s(i, n)$ is not the total information an observer *needs* to predict the state of the destination (this is measured by the single-site entropy $h(i, n)$—see Sect. 3.2.2 and Sect. 4.1.4). It is the total *obtained* by inspecting the sources separately, ignoring any interaction or redundancies.

Note that unlike other information-theoretic variables, we have introduced the *local* separable information before its average, because the quantity is only truly understood it terms of what its local values imply. The average separable information can certainly be defined also:

$$S_X = \langle s_X(n) \rangle_n , \tag{5.5}$$

$$S_X(k) = \langle s_X(n, k) \rangle_n , \tag{5.6}$$

$$S_X = \lim_{k \to \infty} S_X(k). \tag{5.7}$$

Furthermore, we also introduce the notation $S^+_X(k)$ and $S^-_X(k)$ as the averages of positive and negative local values of $s_X(n, k)$ in contributing to the average$S_X(k)$, for example:

$$S_X^+(k) = \langle s_X^+(n,k) \rangle_n, \tag{5.8}$$

$$s_X^+(n,k) = \begin{cases} s_X(n,k) & \text{if } s_X(n,k) \geq 0 \\ 0 & \text{if } s_X(n,k) < 0 \end{cases}. \tag{5.9}$$

Also $S_X^-(k) = \langle s_X^-(n,k) \rangle$ is defined in the opposite manner, and we have:

$$S_X(k) = S_X^+(k) + S_X^-(k). \tag{5.10}$$

$S_X^+(k)$ and $S_X^-(k)$ are used to describe the relative proportions of trivial and non-trivial information modifications in the computations of X (in later chapters).
Finally, for lattice systems we similarly have:

$$S(i) = \langle s(i,n) \rangle_n, \tag{5.11}$$

$$S(i,k) = \langle s(i,n,k) \rangle_n, \tag{5.12}$$

$$S(i) = \lim_{k \to \infty} S(i,k). \tag{5.13}$$

For homogeneous agents we have $S(k) = \langle s(i,n,k) \rangle_{i,n}$, and define S in the limit $k \to \infty$. Also, we can define $S^+(k)$ and $S^-(k)$ for lattice systems as above, and extend the definitions appropriately to the limit $k \to \infty$.

5.2 Local Information Modification in Cellular Automata

Local separable information was applied with $k = 16$ to study information modification in the same sample ECA runs where we analysed information storage and transfer in Sects. 3.3 and 4.2; i.e. for rules 54 (Fig. 3.4a), 110 (Fig. 3.5a), ϕ_{par} (Fig. 3.6a), 18 (Fig. 3.9a), 22 (Fig. 3.10a) and 30 (Fig. 3.11a).
We discuss the key results from this application here:

- Negative values of local separable information provide the first quantitative identification of *hard collisions* between particles as *dominant* non-trivial information modification events in Sect. 5.2.1. This is the case for both regular gliders and domain walls, and we observe a short time delay between the apparent collisions and the identified information modification points.
- That non-trivial information modification events can also occur separately from particle collisions, for example in *soft collisions* between gliders and background domains (Sect. 5.2.2), storage modifications in non-periodic background domains (Sect. 5.2.3), as well as throughout the chaotic dynamics of rules 22 and 30 (Sect. 5.2.4).
- That appropriately large values of past history k are required to provide the perspective of distributed computation and identify non-trivial modification points.

5.2.1 Hard Particle Collisions as Dominant Modification Events

The simple gliders in ECA rule 54 give rise to relatively simple collisions which we focus on in our discussion of $s(i, n, k = 16)$ here (see Fig. 3.4a). Notice that the positive values of $s(i, n, k = 16)$ are concentrated in the domain regions and at the stationary gliders (α and β). As expected, these regions are undertaking trivial computations only. More importantly, the negative values of $s(i, n, k = 16)$ are also shown in Fig. 3.4a, with their positions circled there. The dominant negative values are clearly concentrated around the areas of *collisions between the gliders*, including collisions between the travelling gliders only (marked by "A") and between the travelling gliders and the stationary gliders (marked by "B", "C" and "D"). This clearly **confirms the glider collisions as non-trivial information modification events**. We term them **hard collisions** because they are collisions between explicit emergent structures.

For example, collision "A" involves the γ^+ and γ^- particles interacting to produce a β particle ($\gamma^+ + \gamma^- \rightarrow \beta$ [9]). The only information modification point highlighted is one time step below that at which the gliders initially appear to collide—these points are marked "o" and "$*$" respectively in the close-up of raw states in Fig. 3.8. The periodic pattern from the past of the destination breaks at "$*$", however the neighbouring sources are still able to support separate prediction of the state, i.e.: $a(i, n, k = 16) = -1.09$ bits, $t(i, j = 1, n, k = 16) = 2.02$ bits and $t(i, j = -1, n, k = 16) = 2.02$ bits, giving $s(i, n, k = 16) = 2.95$ bits. This is no longer the case however at "o" where our measure has identified the modification point; there we have $a(i, n, k = 16) = -3.00$ bits, $t(i, j = 1, n, k = 16) = 0.91$ bits and $t(i, j = -1, n, k = 16) = 0.90$ bits, with $s(i, n, k = 16) = -1.19$ bits suggesting a non-trivial information modification.

A *delay* is also observed before the identified information modification points of collision types "B" ($\gamma^+ + \beta \rightarrow \gamma^+$, or vice-versa in γ-types), "C" ($\gamma^- + \alpha \rightarrow \gamma^- + \alpha + 2\gamma^+$, or vice-versa) and "D" ($2\gamma^+ + \alpha + 2\gamma^- \rightarrow \alpha$). Possibly these delays represent a time-lag of information processing. Not surprisingly, the results for these other collision types imply that the information modification points are associated with the *creation of new behaviour*: in "B" and "C" these occur along the newly created γ gliders, and for "C" and "D" in the new α blinkers.

We observe similar results in the profile of $s(i, n, k = 10)$ for ϕ_{par} in Fig. 3.6a, confirming the particle collisions here as non-trivial information modification events. This completes the evidence for all of the conjectures about the role of emergent structures in this human-understandable distributed computation.

The results for $s(i, n, k = 16)$ for ECA rule 110 in Fig. 3.5a are also similar. Here, we have collisions "A" and "B" which show non-trivial information modification points slightly delayed from the collision in a similar fashion to those for rule 54. We note that collisions between some of the more complex glider structures in rule 110 (not shown) exhibit non-trivial information modification points which are more difficult to interpret, and which are even more delayed from the initiation of the collision. The larger delay is perhaps this is a reflection of the more complex gliders requiring more time steps for the processing to take place. An interesting result not seen for rule 54 is a collision where an incident glider is absorbed by a blinker (not shown), without any modification to the absorbing blinker. No information modification is detected for this absorption event by $s(i, n, k = 16)$: this is because the information storage for the absorbing blinker is sufficient to predict the dynamics at this interaction.

Furthermore, as displayed in Fig. 3.9a, **the separable information quite clearly identifies the hard collision between the domain walls as dominant information modification events** for rule 18. The initial information modification event is clearly where one would initially identify the collision point, yet it is followed by two secondary information modification points separated by two time steps. At the raw states of these three collision points in Fig. 3.9a, the outer domains have effectively coalesced (inferred by *spatial* scanning). The source in say the left outer domain (in the context of the past in the inner domain) indicates that the domain walls should intrude into the inner domain; the domain wall is not observed though because the outer domains have coalesced, and this is misinformative. The same applies for the source in the right domain, so the separable information is negative at these points due to the transfer sources. Indeed an observer following the computational perspective and scanning the *temporal* pattern cannot be certain that the new domain has taken hold at this particular cell until observing a "1" at the alternate phase (see discussion of the two phases of the domain in Sect. 3.3.5). As such, these information modification events continue to be observed until a "1" confirms the outer domains have joined.[5] This in some ways parallels the observation of delays in information processing observed earlier.

Importantly, this result provides evidence that collision of irregular particles are information modification events, as expected. It is also worth noting that these collisions always result in the destruction of the domain walls (and the inner domain), indicating that our method captures *destruction-type modification events* as well as creation. (This is also true for the $\gamma^+ + \gamma^- + \beta \rightarrow \oslash$ event in rule 54, not shown).

Interestingly also, note that the glider collisions in rules 54 and 110 always occurred with $t(i, j, n, k) > 0$ for at least one j. In contrast, the domain wall collisions in rule 18 have a different basis for non-trivial modification because we have $t(i, j, n, k) < 0$ for both j while the $a(i, n, k)$ remains positive.

While particle collisions are the *dominant* non-trivial information modification events, they are not the only such events in these dynamics. In the next sections we discuss the manifestation of non-trivial information modification in gliders, non-periodic background domains, and their proliferation in chaotic dynamics.

[5] Only then does the active information storage go negative, because from a temporal perspective the inner domain no longer continues. The separable information is positive here though, as the transfer sources provide positive information about their domains intruding (having coalesced).

5.2.2 Soft Collisions Between Gliders and the Domain

Interestingly, weak non-trivial information modification points continue to be identified at every second point along all the γ^+ and γ^- particles in rule 54 after the initial collisions. These are too weak to appear in Fig. 3.4a but can be seen for a similar glider in rule 110 in Fig. 3.5a. This was unexpected from our earlier hypothesis. However, these events can be understood as non-trivial computations of the *continuation* of the glider in the *absence* of a collision. From another perspective, they are ***soft collisions* of the glider with the periodic structures and ambient transfer in the domain.** Recall the ambient transfer refers to the small but non-zero information transfer in periodic domains indicating the *absence* of gliders (see Sect. 4.2.3). The term soft collisions indicates the qualitative contrast with hard collisions between particles, and that the gliders continue unperturbed. These soft collision events are more significant closer to the hard collisions, since the ambient transfer is stronger in the wake of the gliders that caused these collisions (see Sect. 4.2.3).

Also, note that these soft collision events occur with $a(i, n, k) < 0$ (since the domain is misinformative) and at least one $t(i, j, n, k) > 0$ (since the glider contains strong transfer). Importantly also, *some* soft collision events (e.g. in rule 110) occur with a larger magnitude $s(i, n, k) < 0$ than for some hard collision events. These facts together mean that hard and soft collisions are not differentiable using $s(i, n, k)$ alone, or by examining the underlying values of $a(i, n, k)$ and $t(i, j, n, k)$ that produce these values $s(i, n, k) < 0$. From the perspective of these measurements, *they are both simply occurrences of non-trivial information modification*.In future work, we will investigate methods to formally quantify the difference between these collision types.

In contrast to the gliders, the domain walls in rule 18 appear to give rise to only positive values of $s(i, n, k = 16)$. This indicates that the domains walls contain only trivial information modification, in contrast with regular gliders which required a small amount of non-trivial information processing in order to compute their continuation. This is perhaps akin to the observation in [20] that the domain walls in rule 146 are largely determined by the dynamics on either side, i.e. they are not the result of any interaction per se but of dominance from a single source at each time step.

5.2.3 Storage Modifications in Non-Periodic Domains

Also, as displayed in Fig. 3.9a, the background domain of rule 18 takes values of $s(i, n, k = 16)$ as either positive or negative with $a(i, n, k = 16)$, since $t(i, j = 1, n, k = 16)$ and $t(i, j = -1, n, k = 16)$ vanish at these points. As described in Sect. 3.3.5, $a(i, n, k = 16)$ is positive for the "0" values at every second site, whereas for the alternate sites it is positive where these are "1" but negative where they are "0". This indicates that the "0" sites for every second point and the "1"'s which only occur in the alternate phase are trivial computations dominated by information storage. In contrast then, some minor information processing is required to compute the "0" sites in the alternate phase. The occurrence of a "0" in this phase is a non-trivial information modification because it makes the task of temporally determining the phase more ambiguous in the future. With $a(i, n, k = 16) < 0$ as the determining factor here, it could indeed be viewed as a *storage modification*.

5.2.4 Proliferation of Information Modification in Chaotic Dynamics

We have also applied $s(i, n, k = 16)$ to ECA rule 22, as displayed in Fig. 3.10a, and rule 30 in Fig. 3.11a. The profiles contain many points of both positive and negative local separable information. Indeed the presence of negative values implies the occurrence of non-trivial information modification, yet there does not appear to be any structure to these profiles. Again, this aligns well with the lack of coherent structure observed in the local information storage and transfer profiles for these rules in Sects. 3.3.8 and 4.2.2, and from the local statistical complexity profile of rule 22 [20].

In both rules 22 and 30, we observe non-trivial information modifications with both $a(i, n, k) < 0$ and $t(i, j, n, k) < 0$ for all j, *in addition* to the other sign combinations for these events previously observed in rules 110, 54 and 18 (see summary in Table 5.2). Indeed these events, along with those with $a(i, n, k) > 0$ and $t(i, j, n, k) < 0$ for all j, are the most prevalent and strongest non-trivial modification events for rule 22. We cannot conclude that these types of modifications events are prohibited between coherent structures though: at this stage we have no theoretical basis for such a conclusion, and indeed such modifications between coherent structures may exist in other examples. Further investigation is required on this topic. Nonetheless, the result itself is important since it demonstrates that non-trivial information modifications can occur with all sign combinations of $a(i, n, k)$, $t(i, j = 1, n, k)$ and $t(i, j = -1, n, k)$ (except for all of them being positive, since this would leave $s(i, n, k) > 0$).

5.2.5 Modification Only Understood in Context of Past History

Finally, we note that measurements of $s(i, n, k)$ must be performed with an appropriately large value of k. In Sect. 3.3.2 and 4.2.2, we observed that for appropriate measurement of information storage and transfer k should be selected to be as large as possible for accuracy, at least larger than the scale of the period of the regular background domain for CA filtering purposes. Indeed, using sufficiently large values of k provides the perspective of distributed computation, by properly separating information storage and transfer. Since the definition of $s(i, n, k)$ is dependent on this perspective of distributed computation, and accurate measurement of information storage and transfer, the preceding text has assumed the same applies here. In testing this assumption, we note that for rule 54 $k < 4$ could not distinguish any collision points clearly from the domains and particles, and even $k < 8$ could not distinguish *all* of them (results not shown). Correct quantification of separable information requires satisfactory estimates of information storage and transfer, and accurate distinction between the two. The key result here is that **detecting information modification requires the perspective of distributed computation, which is highly dependent on establishing the context of the past state of the destination**.

5.3 Irreversibly Destroyed Information

Questions remain, even from a qualitative perspective, of the relationship between the concepts of information modification and *irreversible information destruction*. Logical irreversibility is a fundamentally important concept because it provides strong ties between information theory and thermodynamics. This is primarily through Landauer's principle [21], which states that irreversible destruction of one bit of information results in dissipation of at least $kT \ln 2$ J of energy[6] into the environment (i.e. an entropy increase in the environment by this amount) [22–24].[7]

The term information *modification*, and indeed the focus in the preceding discussion of perturbation of emergent structures, certainly both allude to information destruction. This inevitably raises the question of whether the concepts of information modification and irreversible information destruction are indeed the same. If not, then how should one measure irreversible information destruction in a *distributed computation*, in particular on a local scale? If this can be done, what then can we learn about the relationship between the concepts of information modification and irreversible information destruction?

We hypothesise that the concepts are complementary but distinct. They are likely to overlap in some instances. Certainly most particle collisions involve destruction of at least one incoming particle. Also, since the resulting emergent dynamics of some particle collisions are possible from other starting configurations,[8] then information is destroyed in the collision about what the starting configuration was. However there are likely to be differences between the two concepts, e.g. information could be destroyed without any interaction between information sources, and certain collision events could be completely reversible or else destroy comparatively little information.

In Sect. 5.3.1 we discuss how to quantify irreversible information destruction on a local scale in distributed computation. The perspective of distributed computation is

[6] T is the absolute temperature and k is Boltzmann's constant.

[7] Maroney [25] argues that while a logically irreversible transformation of information does generate this amount of heat, it can in fact be accomplished by a *thermodynamically reversible* mechanism.

[8] For example, the result of an annihilation could be trivially reproduced without any of the particles existing in the first place.

critical here: it means we wish to quantify information destruction on a local scale in space and time, which is an extension over previous considerations in time only. We then apply the measure to our CA examples in Sect. 5.3.2, which is a novel application since consideration of CAs in this context has typically focussed on whether rules are reversible [26–29], but not on the specific locations in space and time where information is destroyed and how this relates to the dynamics of computation. We contrast the findings with our results for non-trivial information modification from the separable information. As expected, we demonstrate that the two concepts are complementary but distinct: certainly they overlap in many particle collision cases, but we present instances of each occurring in isolation.

5.3.1 Measuring Information Destruction in Distributed Computation

An operation is defined to be *logically irreversible* if the output does not uniquely define the inputs [21]. The classic example of this concept is the "reset to zero" (RTZ) operation [24, 25, 30]: $f(x) = 0, x \in \{0, 1\}$. Clearly here the output 0 cannot be used to recover information about what the input bit x was, so the operation is irreversible and information is *destroyed*. This simple example demonstrates that computational processes that deterministically map inputs to an output do not necessarily map the same output uniquely back to its input. The RTZ operation is generally juxtaposed with the bit flipping operation: $f(0) = 1, f(1) = 0$. One can clearly recover the input from the output of a bit flip, so information is preserved in that operation.

The amount of *irreversibly destroyed information* can be captured information-theoretically as the *uncertainty* in the inputs given the outputs (in alignment with [31]). For a univariate process $X \rightarrow X'$ we can write the average amount of **information destruction** at each time step as:

$$D_X = H(X \mid X'), \tag{5.14}$$

For the RTZ operation, we have $D_X = H(X)$ (since there is no information in the output X'); so for a maximum entropy input distribution, one bit of information is destroyed in the operation.[9] In the following, we build an approach to measuring local information destruction in distributed computation. We start with measuring information destruction on a local scale in time in univariate processes, then in multivariate processes, and finally discuss the local scale in space and time in multivariate processes.

---

[9] With this formulation we see that the approach to information processing in [3] (referred to at the start of this chapter) has aspects more akin to destroying redundant information than information modification. Indeed the authors describe part of the approach as measuring "minimization of spurious information" retained in the output.

5.3.1.1 Local Information Destruction in Time-Series Processes

In extending this notion to distributed computation, we first need to consider ongoing univariate time-series processes where $x_n \rightarrow x_{n+1}$ for each time step n. We can write the **local information destruction** at each time step n of process X as $d_X(n+1)$:

$$D_X = \langle d_x(n+1) \rangle_n , \tag{5.15}$$

$$d_X(n+1) = h(x_n \mid x_{n+1}). \tag{5.16}$$

Additionally though, where the future state x_{n+1} is not statistically independent of multiple *past* states $\{x_{n-c} \mid k \geq c > 0\}$ given the previous state x_n,[10] the expression needs to be altered. This is because such statistical dependence means that conversely information about x_n could be conserved beyond the next state x_{n+1} even if such information is not contained in that next state. We are interested in the uncertainty in (or information destroyed about) x_n given all of the *future* states $x^{(k)}$ that information about it could be conserved in, so we write:

$$D_X(k) = H(X \mid X^{(k^+)}), \tag{5.17}$$

$$d_X(n+1,k) = h(x_n \mid x_{n+1}^{(k^+)}). \tag{5.18}$$

Of course, this in general requires the limit $k \rightarrow \infty$ unless the limit of such statistical dependence can be established (e.g. with the synchronisation time [32]). Also, the incorporation of k consecutive values here means that we are looking at preservation of information in *embedding vectors* [33], which capture the underlying *state* of the process. This is in alignment with the examination of information destruction in underlying *causal states* of ϵ-machines in [31] (and similarly in [34]), where it is demonstrated that the difference between the statistical complexity and excess entropy is equivalent to the irreversibly destroyed information about the past state of each time step.[11]

5.3.1.2 Local Information Destruction in Time in Multivariate Systems

When considering multivariate systems with a joint vector of states $\mathbf{X}$ where at each time step $\mathbf{x}_n \rightarrow \mathbf{x}_{n+1}$, the *information destruction in multivariate systems* can be

[10] This can be because these past states are direct causal sources of the future state, or are indirectly causal via other variables (such as neighbouring cells in a CA, i.e. as per Sect. 3.1), or perhaps both the past states and future state have a common causal driver.

[11] Note the important distinction here. The statistical complexity—excess entropy interpretation computes the information that will be destroyed *in total* about the current underlying ϵ-machine state, over an arbitrary number of time steps. The interpretation in Eq. (5.18) computes the information destroyed about the current underlying state *at the next state transition* $x_n^{(k^+)} \rightarrow x_{n+1}^{(k^+)}$ only. The distinction parallels that between the excess entropy and active information storage discussed in Chap. 3, and experimental results from the two views could be contrasted in future work.

written on average and locally in time as:

$$D_{\mathbf{X}}(k) = H(\mathbf{X} \mid \mathbf{X}^{(k^+)}), \tag{5.19}$$

$$d_{\mathbf{X}}(n+1, k) = h(\mathbf{x}_n \mid \mathbf{x}_{n+1}^{(k^+)}). \tag{5.20}$$

However, because we are considering the system $\mathbf{X}$ *as a whole* then under certain assumptions (including *causal closure* of the system)[12] we can simplify this back to:

$$D_{\mathbf{X}} = H(\mathbf{X} \mid \mathbf{X}'), \tag{5.21}$$

$$d_{\mathbf{X}}(n+1) = h(\mathbf{x}_n \mid \mathbf{x}_{n+1}). \tag{5.22}$$

An important interpretation is Bennett's description [22] of "merging of two computational paths" as logically irreversible. The computational paths referred to here are trajectories through the *state-space* of the joint system (which we have already encountered for CAs [37] in Sect. 2.3). For a deterministic system, any global state is only mapped to a single next state, yet there may be multiple precursor states leading to one next state. This is the merging of computational paths: if at time step $n+1$ we reach a global system state $\mathbf{x}_{n+1}$ with multiple precursor states $\mathbf{x}_n$, then we have non-zero irreversibly destroyed information $d_{\mathbf{X}}(n+1) = h(\mathbf{x}_n \mid \mathbf{x}_{n+1})$ at time step $n+1$. This means that a (finite) discrete dynamical system is reversible if and only if its state-space contains states on attractor cycles only without any transient states leading up to them [26]. In Appendix F we discuss the fact that computational paths cannot merge and information cannot be destroyed in *thermodynamically closed systems*; we can only measure information destruction in *open computational systems*, and what we measure is a departure of information from this system into the external, unobserved environment.

5.3.1.3 Local Information Destruction in Time and Space in Multivariate Systems

Measuring $D_{\mathbf{X}} = H(\mathbf{X} \mid \mathbf{X}')$ has obvious difficulties for large systems, and while localising with $d_{\mathbf{X}}(n+1)$ tells us when in time information was destroyed, it does not tell us where in space this occurred.

To address this question, we introduce the **local information destruction in space and time in multivariate systems**. This quantity, $d_{X_i}(n+1)$ or $d(i, n+1)$, measures

[12] *Assumption 1*: that all causal inputs to the computation are considered to be within the system, i.e. it is causally closed or closed to efficient cause [35] (which is stronger than informational closure [36]). *Assumption 2*: (the normal case where) only the state at time step n is a direct causal input to the state at time step $n+1$. Under these assumptions, none of the conditions described in footnote 10 apply to the system as a whole, and so the future state $\mathbf{x}_{n+1}$ is statistically independent of multiple past states $\{\mathbf{x}_{n-c} \mid k \geq c > 0\}$ given the previous state $\mathbf{x}_n$ (i.e. $I(\mathbf{x}_{n-c}; \mathbf{x}_{n+1} \mid \mathbf{x}_n) = 0, \forall k \geq c > 0$). Note that our assumption of causal closure allows for stochasticity in the computation of the next state of the system, but not for correlations across time in such stochasticity.

the amount of uncertainty in (or information destroyed about) the previous state $x_{i,n}$ of variable X_i given the next state $\mathbf{x}_{n+1}$ of the system $\mathbf{X}$ at time $n+1$:

$$d(i, n+1) = h(x_{i,n} \mid \mathbf{x}_{n+1}), \tag{5.23}$$
$$D(i) = H(X_i \mid \mathbf{X}), \tag{5.24}$$
$$= \langle d(i, n+1) \rangle_n \tag{5.25}$$

In comparison to a single time-series Eq. (5.16), we try to find information about x_n in the whole system $\mathbf{x}_{n+1}$ rather than only in x_{n+1}. This is because in the context of the *distributed computation* we only consider the information destroyed if it is no longer available *anywhere* in the system. In a CA for example, $x_{i,n}$ has a direct causal effect on its future light-cone, so these agents can directly contain information about $x_{i,n}$. Yet extra information about $x_{i,n}$ may exist outside of this future light-cone, for example where there are longer-range correlations in the system. From another perspective, the extended system $\mathbf{x}_{n+1}$ also contains information about the neighbours of X_i at time n (and their own neighbours, and so on), and that information can be helpful in decoding $x_{i,n}$ based on the next states $\mathbf{x}_{n+1}$ they produced together.[13] As a concrete demonstration, it is known that "the inverse map of a reversible CA is always itself a CA, but the inverse CA does not necessarily have the same radii" or neighbourhood r ([29] citing [27, 28]). Without knowing the limits within which to recover this information therefore, in general one should examine the whole system $\mathbf{x}_{n+1}$. While we are locally identifying where the information may have been destroyed from, certainly there is a "departure from locality" [38] in recognising where that information may still exist.

Furthermore, we note that $d_{\mathbf{X}}(n+1) \neq \sum_i d(i, n+1)$ in general. While $d(i, n+1)$ certainly quantifies the amount of information destroyed about $x_{i,n}$, this may include redundancies with information destroyed about say $x_{i-j,n}$. Correct summing to obtain $d_{\mathbf{X}}(n+1)$ would involve incrementally conditioning on the destroyed information from previously considered agents (in the style of Eq. (4.47)).

In practice, the number of available observations normally precludes the use of the whole system state $\mathbf{x}_{n+1}$. As such, one needs to restrict analysis to a *subset* of the next state of the system. The most logical restriction is to agents which have shorter paths of causal links to x_n within its past light-cone [39, 40]. In lattice systems with local interactions (e.g. CAs) one can take the subset $\mathbf{X}_{i,m}$ of agents within m cells of X_i, and so express this **local information destruction approximation** as:

$$d(i, n+1, m) = h(x_{i,n} \mid \mathbf{x}_{i,n+1,m}), \tag{5.26}$$
$$\mathbf{x}_{i,n+1,m} = \{x_{i+q,n+1} \mid \forall q : -m \leq q \leq +m\}. \tag{5.27}$$

Figure 5.3 displays a diagram of $d(i, n+1, m)$. Obviously, $d(i, n+1, m)$ converges with $d(i, n+1)$ in the limit as m envelopes the whole system.

---

[13] This is akin to using the next state of the whole system $\mathbf{x}_{n+1}$ to compute pre-images of the whole system $\mathbf{x}_n$, by gradually building the previous state cell by cell [37].

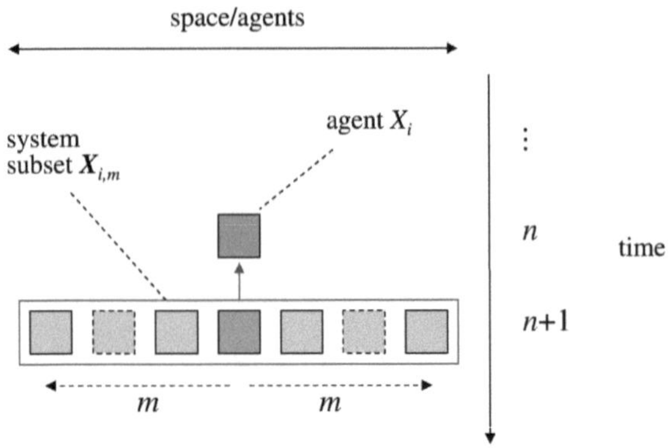

Fig. 5.3 Local information destruction (approximation) $d(i, n+1, m)$ in lattice systems: information in the previous state $x_{i,n}$ of X_i that is no longer contained in the subset $\mathbf{x}_{i,n+1,m}$ of the next state of the system $\mathbf{X}_i$. The subset refers to agents within m cells of X_i

Finally, in comparison to Eq. (5.18) note that we have not included k steps of $\mathbf{x}_{i,n+1}$ into the future in $d(i, n+1)$ in Eq. (5.23). Under the assumptions described in footnote 12, $x_{i,n}$ is conditionally independent of these future states given $\mathbf{x}_{i,n+1}$, so this is not necessary. Where those assumptions are not valid, these k steps should theoretically be included in Eq. (5.23). For the approximation Eq. (5.26), we note that the *restriction* to $\mathbf{x}_{i,n+1,m}$ may itself result in again rendering $x_{i,n}$ conditionally dependent on the future states given $\mathbf{x}_{i,n+1,m}$. However, given that the approximation $d(i, n+1, m)$ was introduced to counter a limited number of available approximations, there is little point in re-burdening the measure by including these k steps.

5.3.2 *Irreversible Information Destruction in Cellular Automata*

Local information destruction $d(i, n, m = 8)$ was applied to the same sample ECA runs where information storage, transfer and modification was analysed in Sects. 3.3, 4.2 and 5.2. The results are displayed for rules 54 in Fig. 5.4 and 18 in Fig. 5.5 as illustrative examples, with their results contrasted with the separable information profiles in those figures.

We demonstrate that as expected the dominant information destruction events are particle collisions, however not all particle collisions involve significant information destruction. Also, signification information destruction can be associated with events other than particle collisions. The concepts of information modification and irreversible information destruction are therefore demonstrated to be distinct but complementary.

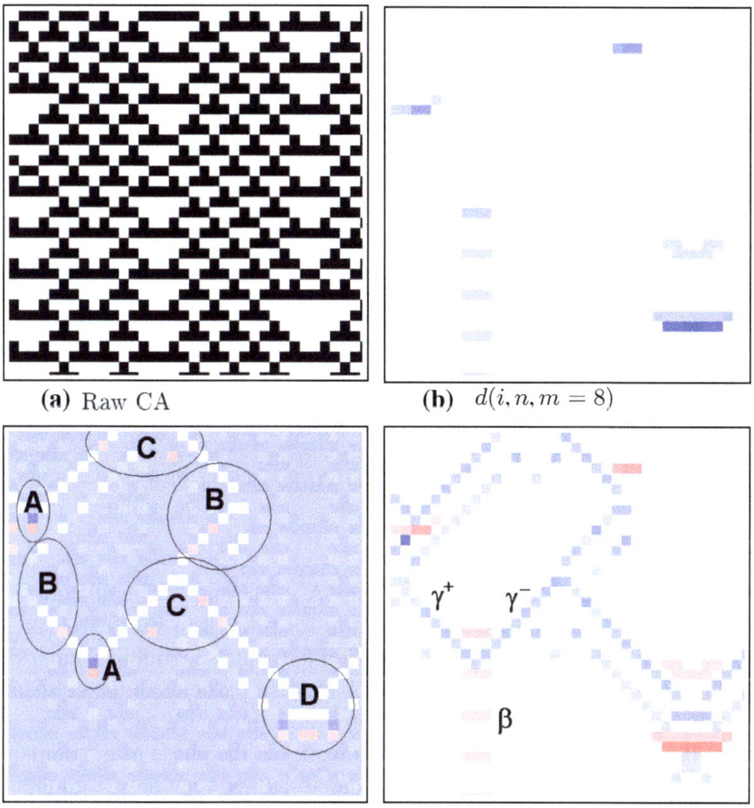

(a) Raw CA **(b)** $d(i, n, m = 8)$

(c) $s(i, n, k = 16)$ **(d)** $d(i, n, m = 8)$ against $h_\mu(i, n, k = 16)$

Fig. 5.4 **a** Local information destruction and modification in rule 54 (35 time steps displayed for 35 cells). Cells are coloured *blue* for positive values and *red* for negative in (**b**) and (**c**). **b** Local information destruction $d(i, n, m = 8)$ with max. 12.53 bits, min. 0.00 bits; **c** Local separable information with collisions marked—copied from Fig. 3.4a; **d** Local information destruction $d(i, n, m = 8)$ from (**b**) plotted in red against local temporal entropy rate $h_\mu(i, n, k = 16)$ from Fig. 3.4d. (NB: Fig. 5.4**a** and **c** Reprinted with permission from J. T. Lizier et al. [16] Copyright 2010, American Institute of Physics)

5.3.2.1 Large Information Destruction Associated With Many Particle Collisions

As expected, **the dominant information destruction events are associated with particle collisions**. This is shown for both periodic gliders in rule 54 (and 110 and ϕ_{par}, not shown) and in domain walls in rule 18. This result underlines similarities between information modification and information destruction.

In contrast with non-trivial information modification events however, the (initial) information destruction associated with particle collisions appears to occur *before*

(a) Raw CA (b) $d(i, n, m = 8)$ (c) $s(i, n, k = 16)$

Fig. 5.5 **a** Local information destruction and modification in rule 18 (67 time steps displayed for 67 cells). Cells are coloured *blue* for positive values and *red* for negative in (**b**) and (**c**). **b** Local information destruction $d(i, n, m = 8)$ with collision marked and max. 12.39 bits, min. 0.00 bits; **c** Local separable information with collision marked—copied from Fig. 3.9a. (NB: Fig. 5.5a and c Reprinted with permission from J. T. Lizier et al. [16] Copyright 2010, American Institute of Physics)

the collision itself. This is the case for the "A", "B" and "D" collisions for rule 54 in Fig. 5.4d and the domain wall collision for rule 18 in Fig. 5.5b. The spatial location of such early information destruction is typically within an inner domain that will be *destroyed* by being sandwiched between particles in the collision. The temporal location is where that inner domain is no longer wide enough for its spatial pattern to be distinguished properly (e.g. to work out the phase of the domain)—as such, the information about it can be considered to have been destroyed there. This example explicitly shows the manner in which the local information destruction takes a *spatial perspective* of patterns (from examining the spatial array $\mathbf{x}_{i,n+1,m}$). On the other hand, analysis of distributed computation takes the *temporal perspective* of the history of each variable and so the information modification is not realised until later in time. We could also interpret the distinction in the temporal location in that analysis of computation looks at how computations unfold going forward in time, while analysis of information destruction looks backwards in time.

As an example here, collision "A" ($\gamma^+ + \gamma^- \rightarrow \beta$) in Fig. 5.4d shows that the (initial) information destruction occurs three time steps prior to the point marked "$*$" in Fig. 3.8. At the middle information destruction point, we have $p(x_{i,n} \mid \mathbf{x}_{i,n+1,m=8}) = 0.40$ giving $h(x_{i,n} \mid \mathbf{x}_{i,n+1,m=8}) = 1.34$ bits, so there is significant uncertainty left in what the previous state $x_{i,n}$ was.

In a similar fashion, for the domain wall annihilation of rule 18 in Fig. 5.5b information is destroyed (circled) even further in advance of what is determined to be the information modification point. At these points an observer of the *spatial* system sees only a single domain, and can no longer tell whether that was the case previously or whether a particle annihilation just occurred.[14] Indeed, the information destruction

[14] Note the similarly large local information destruction in the unperturbed domain wall to the left of the particle annihilation. These values are an artifact of $m = 8$ not being large enough to see

is large because the particle annihilation that actually occurred was a comparatively rare event. Again, due to the temporal perspective of computation there is a delay before the local information modification is realised.

Finally, we note that large information destruction was also associated with the particle collision events in the human-understandable computation of ϕ_{par} (not shown).

5.3.2.2 Information Modification Without Destruction

It is possible though to have information modification events without significant information destruction. A prime example here is collision "C" in rule 54 (see Fig. 5.4d). Here, there is vanishing[15] information destruction associated with the glider collision $\gamma^- + \alpha \rightarrow \gamma^- + \alpha + 2\gamma^+$, or vice-versa. This is because the products of this collision are not produced via any other common particle interactions. As such, these modification events are effectively reversible.

5.3.2.3 Information Destruction Without Modification

Correspondingly, **we also observe information destruction events without any non-trivial modification occurring**. The most prominent example here are the repetitive destructions of approximately 0.5 bits along the β blinker in the bottom of Fig. 5.4b after its creation in collision "A". Here, information is destroyed because an observer is uncertain whether the blinker was just created from a collision A event ($\gamma^+ + \gamma^- \rightarrow \beta$) or whether the blinker existed previously. Since collision "A" events are relatively common, there is a large amount of uncertainty at these points.

The periodic repetition of information destruction here parallels that associated with periodic system attractors. Where such attractors have transient states leading up to them, there is at least one state on the attractor with multiple precursors: the preceding state on the attractor and the incoming transient path(s). As discussed in Sect. 5.3.1.2, there is irreversible information destruction where these computational paths merge since uncertainty remains over whether the system just entered or had already been on the periodic attractor. Interestingly, the information will periodically be destroyed each time the system loops around the attractor. This is not a physical violation: because the system is open there is no restriction on it periodically dumping information and energy into the outside world (akin to an open heat cycle from classical thermodynamics [41]). The β blinkers here could be thought of as a *local*

(Footnote 14 continued)
beyond the large number of consecutive "0" states in the white triangle in Fig. 5.5a to gauge the phase of the surrounding domains.

15 There are a few points around the collision with small non-zero $d(i, n, m = 8)$ up to 0.07 bits. These appear to be largely artifacts of the surrounding area of the system ($m = 8$) analysed, partially due to the occurrence of similar next state configurations with different preceding states in rare dynamics, rather than being due to irreversibility in the collision itself.

periodic attractor, where information is periodically destroyed regarding whether the blinker was just created or had previously existed.

In comparison to the β blinkers, these periodic events occur only with a very small magnitude (<0.03 bits, not visible in Fig. 5.4b) along the α blinkers in rule 54. These blinkers are produced in collision "C" with other products (see above), and in collision "D" ($2\gamma^+ + \alpha + 2\gamma^- \rightarrow \alpha$) where they are produced in isolation. As such, when observing (the relevant phase of) an α blinker as an *isolated* emergent structure in $d(i, n, m = 8)$, an observer is uncertain whether it was just created from collision "D" or had already existed. Since collision "D" is a comparatively rare event though, there is a large amount of destroyed information in the collision itself (on the order of 5 to 7 bits at various local points) but very little in the blinker's periodic continuation. Thus, the trajectory of α is almost completely reversible.

Importantly, from a computational perspective both blinkers and system attractors are purely information storage processes—there is no information modification taking place there. We see that information modification and destruction are complementary: they are similar in being predominantly associated with particle collisions, yet have a number of subtle distinctions described here and above.

5.3.2.4 Other Information Destruction Processes

There are a number of other interesting local information destruction processes highlighted by $d(i, n, m = 8)$ here. In the background domain of rule 18 in Fig. 5.5b, we measure approximately 1 bit of uncertainty regarding whether each preceding site in the alternate phase was a "0" or "1" (see Sects. 3.3.5 and 4.2.4.2 for a description of the phases in the domain of rule 18). The collective next state is useful in decoding the phase of those sites, but cannot necessarily decode the exact values. With the large amount of interaction occurring in the domain via XOR operations (see Sect. 4.2.4.2 and [42]), there are multiple possibilities for the preceding state. Again, this is in contrast to the separable information which in Sect. 5.2.3 measured different levels of information modification in the computation of "1"s and "0"s respectively in this phase.

Furthermore, we note contrasting areas of zero and large information destruction surrounding the domain walls for rule 18. That a domain wall is contained in the past of $\mathbf{x}_{i,n+1,m}$ is generally clear from $\mathbf{x}_{i,n+1,m}$ itself where an observer can see that two phases are out of alignment. For some time steps n along the domain wall, it is clear where the wall lies (e.g. if we have two spatially consecutive "1" sites) and what the previous state must have been. This is because knowing where the wall is provides a key into decoding the sequence of XOR inputs from the previous state. Indeed, using larger values of m would allow this decoding to proceed much further spatially along this time step, spatially extending the regions of zero information destruction. For other time steps however the exact location of the wall remains unclear, due to say several spatially consecutive "0" states across the domain wall; as such, the previous state also remains unclear. This provides an interesting contrast with the temporal entropy rate, which clearly

Table 5.1 Local measures relevant to information modification

Measure	Local definition	Equation no.
Separable information	$s_X(n,k) = a_X(n,k) + \sum_{Y \in \mathbf{V}_X \setminus X} t_{Y \rightarrow X}(n,k)$	Eq. (5.2)
Information destruction	$d(i, n+1, m) = h(x_{i,n} \mid \mathbf{x}_{i,n+1,m})$	Eq. (5.26)

identifies locations of the domain wall due to its temporal perspective in considering an advancing computation.

Indeed, there are contrasting findings of zero and significant information destruction in regular gliders also. The γ gliders of rule 54 in Fig. 5.4b are generally reversible, however the more complex gliders of rule 110 (not shown) typically contain more information destruction in their trajectories.

Finally, we note that rule 22 produces incoherent patterns of information destruction (results not shown), in parallel with the other local information profiles produced for it here.

5.4 Summary

We have introduced the local separable information in Sect. 5.1 to quantify information modification at each spatiotemporal point in a complex system (see definition in Table 5.1). Importantly, the measure describes the manner in which information storage and transfer interact to produce non-trivial computation where "the whole is greater than the sum of the parts". Information modification events occur where the separable information is negative, indicating that separate or independent inspection of the causal information sources is misleading because of non-trivial interaction between these sources. In Sect. 5.2 the local separable information was demonstrated to provide the first quantitative evidence that particle collisions in CAs are the dominant information modification events therein. The measure is capable of identifying events involving both creation and destruction, and interestingly the location of an information modification event often appears delayed perhaps due to a time-lag in information processing. Also, the measure identified a number of other non-trivial information modification events, the properties of which are summarised in Table 5.2. Furthermore, in order to separate information storage and transfer and properly identify information modification, the measure required appropriately long values of past history k in establishing the context of the destination's past and taking the perspective of distributed computation.

We also introduced the local information destruction in Sect. 5.3.1 to quantify the concept of irreversible information destruction on a local scale in space and time. We then contrasted this concept with information modification in CAs in Sect. 5.3.2. As expected, particle collisions were the dominant information destruction events, however we demonstrated that information modification can occur without information destruction and correspondingly information destruction can occur without

Table 5.2 Examples of non-trivial information modification events (i.e. with $s(i, n) < 0$) in cellular automata with specific information storage and transfer properties (after [16])

	All $t(i,j,n) < 0$	Mix of $t(i,j,n)$ signs	All $t(i,j,n) > 0$
$a(i,n) > 0$	Domain wall hard collisions in rule 18 (Section 5.2.1) Most prevalent and strongest events in rule 22 (Section 5.2.4)	Relatively few and weakest events in rule 22 (Section 5.2.4) Hard glider collisions in rules 54 and 110 (Section 5.2.1)	N/A
$a(i,n) < 0$		Soft collisions in rules 54 and 110 (Section 5.2.2)	
	Storage modifications in background of rule 18 (Section 5.2.3)		

information modification. The complimentary nature of the concepts was further underlined in that where they both occur in particle collisions, information destruction is typically identified prior to the modification. We also emphasised that local analysis of information destruction takes a spatial perspective of emergent patterns, in contrast to the temporal perspective of distributed computation.

This presentation of the separable information to measure non-trivial information modification completes our framework of the fundamental operations of distributed computation. **Together, the measures of the framework have provided the first quantitative evidence for all of the conjectures about the role of emergent structures in distributed computation in CAs**: that blinkers implement information storage, particles are information transfer agents, and particle collisions are information modification events. The framework is unique in relating these three operations of computation, and in providing such evidence for our qualitative understanding of their embodiment. With the framework in place, in the subsequent chapters we will apply these measures to study various systems, and explore what they can tell us about the nature of complex computation.

References

1. O. Kinouchi, M. Copelli, Optimal dynamical range of excitable networks at criticality. Nat. Phys. **2**(5), 348–351 (2006)
2. J.J. Atick, Could information theory provide an ecological theory of sensory processing? Netw. Comput. Neural Syst. **3**(2), 213 (1992)
3. M.A.Sánchez-Montañés, F.J. Corbacho, Towards a new information processing measure for neural computation, ed. by J. Dorronsoroin. *Proceedings of the International Conference on Artificial Neural Networks (ICANN 2002)*, Madrid, Spain. ser. Lecture Notes in Computer Science, vol. 2415 (Berlin/Heidelberg: Springer-Verlag, 2002), pp. 637–642

4. T. Yamada, K. Aihara, Spatio-temporal complex dynamics and computation in chaotic neural networks, in *Proceedings of the IEEE Symposium on Emerging Technologies and Factory Automation (ETFA '94)*, Tokyo. IEEE, 1994, pp. 239–244
5. M.H. Jakubowski, K. Steiglitz, R. Squier, Information transfer between solitary waves in the saturable Schrödinger equation. Phys. Rev. E **56**(6), 7267 (1997)
6. A. Adamatzky (ed.), *Collision-Based Computing* (Springer-Verlag, Berlin, 2002)
7. D.E. Edmundson, R.H. Enns, Fully 3-dimensional collisions of bistable light bullets. Opt. Lett. **18**, 1609–1611 (1993)
8. C.G. Langton, Computation at the edge of chaos: phase transitions and emergent computation. Physica D **42**(1–3), 12–37 (1990)
9. W. Hordijk, C.R. Shalizi, J.P. Crutchfield, Upper bound on the products of particle interactions in cellular automata. Physica D **154**(3–4), 240–258 (2001)
10. N. Boccara, J. Nasser, M. Roger, Particlelike structures and their interactions in spatiotemporal patterns generated by one-dimensional deterministic cellular-automaton rules. Phys. Rev. A **44**(2), 866–875 (1991)
11. B. Martin, A group interpretation of particles generated by one dimensional cellular automaton, Wolfram's 54 rule. Int. J. Mod. Phys. C **11**(1), 101–123 (2000)
12. G.J. Martinez, A. Adamatzky, H.V. McIntosh, Phenomenology of glider collisions in cellular automaton rule 54 and associated logical gates. Chaos, Solitons Fractals **28**(1), 100–111 (2006)
13. M. Mitchell, J.P. Crutchfield, P.T. Hraber, Evolving cellular automata to perform computations: mechanisms and impediments. Physica D **75**, 361–391 (1994)
14. M. Mitchell, J.P. Crutchfield, R. Das, Evolving cellular automata with genetic algorithms: a review of recent work, ed. by E.D. Goodman, W. Punch, V. Uskov. in *Proceedings of the First International Conference on Evolutionary Computation and Its Applications*, Moscow, Russia: Russian Academy of Sciences, 1996
15. J.T. Lizier, M. Prokopenko, A.Y. Zomaya, Detecting non-trivial computation in complex dynamics, ed. by F. Almeida e Costa, L.M. Rocha, E. Costa, I. Harvey, A. Coutinho, in *Proceedings of the 9th European Conference on Artificial Life (ECAL)*, Lisbon, Portugal, ser. Lecture Notes in Artificial Intelligence, vol. 4648. (Springer, Berlin/Heidelberg, 2007), pp. 895–904
16. J.T. Lizier, M. Prokopenko, A.Y. Zomaya, Information modification and particle collisions in distributed computation. Chaos, **20**(3), 037109 (2010)
17. N. Lüdtke, S. Panzeri, M. Brown, D.S. Broomhead, J. Knowles, M.A. Montemurro, D.B. Kell, Information-theoretic sensitivity analysis: a general method for credit assignment in complex networks. J. Roy. Soc. Interface **5**(19), 223–235 (2008)
18. M. Prokopenko, F. Boschietti, A.J. Ryan, An information theoretic primer on complexity, self organization, and emergence. Complexity **15**(1), 11–28 (2009)
19. D.G. Green, Emergent behavior in biological systems. Complex. Int. **1**, (1994), paper ID: green01
20. C.R. Shalizi, R. Haslinger, J.-B. Rouquier, K.L. Klinkner, C. Moore, Automatic filters for the detection of coherent structure in spatiotemporal systems. Phys. Rev. E **73**(3), 036104 (2006)
21. R. Landauer, Irreversibility and heat generation in the computing process. IBM J. Res. Devel. **5**, 183–191 (1961)
22. C.H. Bennett, Notes on Landauer's principle, reversible computation, and Maxwell's Demon. Stud. Hist. Philos. Sci. Part B **34**(3), 501–510 (2003)
23. B. Piechocinska, Information erasure. Phys. Rev. A **61**(6), 062314 (2000)
24. S. Lloyd, *Programming the Universe* (Vintage Books, New York, 2006)
25. O.J.E. Maroney, Generalizing Landauer's principle. Phys. Rev. E **79**(3), 031105 (2009)
26. A.W. Burks, On backwards-deterministic, erasable, and Garden-of-Eden automata, Computer and Communication Sciences Department, The University of Michigan, Technical Report 012520–4-T, 1971
27. D. Richardson, Tessellations with local transformations. J. Comput. Syst. Sci. **6**(5), 373–388 (1972)
28. T. Toffoli, N.H. Margolus, Invertible cellular automata: a review. Physica D **45**(1–3), 229–253 (1990)

29. T. Helvik, K. Lindgren, M.G. Nordahl, Continuity of information transport in surjective cellular automata. Commun. Math. Phys. **272**(1), 53–74 (2007)
30. C. Seife, *Decoding the universe* (Penguin Group, New York, 2006)
31. K. Wiesner, M. Gu, E. Rieper, V. Vedral, Information erasure lurking behind measures of complexity, 2009, arXiv:0905.2918v1. Available: http://www.arxiv.org/abs/0905.2918
32. D.P. Feldman, J.P. Crutchfield, Synchronizing to periodicity: the transient information and synchronization time of periodic sequences. Adv. Complex Syst. **7**(3–4), 329–355 (2004)
33. F. Takens, in *Detecting strange attractors in turbulence*, ed. by D. Rand, L.-S. Young. Dynamical Systems and Turbulence, Warwick, ser. Lecture Notes in Mathematics, (Springer, Berlin/Heidelberg, 1981), pp. 366–381 (1980)
34. J.R. Mahoney, C.J. Ellison, J.P. Crutchfield, Information accessibility and cryptic processes. J. Phys. A **42**(36), 362002 (2009)
35. R. Rosen, *Life Itself: A Comprehensive Enquiry into the Nature, Origin and Fabrication of Life* (Columbia University Press, New York, 1991)
36. N. Bertschinger, E. Olbrich, N. Ay, J. Jost, Information and closure in systems theory, ed. by S. Artmann, P. Dittrich in *Proceedings of the 7th German Workshop on Artificial Life (GWAL-7)*, Jena, Germany (IOS Press, Amsterdam, 2006)
37. A. Wuensche, Classifying cellular automata automatically: finding gliders, filtering, and relating space-time patterns, attractor basins, and the Z parameter. Complexity **4**(3), 47–66 (1999)
38. B. Misra, I. Prigogine, Irreversibility and nonlocality. Lett. Math. Phys. **7**(5), 421–429 (1983)
39. C.R. Shalizi, *Causal architecture, complexity and self-organization in time series and cellular automata*, Ph.D. dissertation, University of Wisconsin-Madison, 2001
40. C.R. Shalizi, K.L. Shalizi, R. Haslinger, Quantifying self-organization with optimal predictors. Phys. Rev. Lett. **93**(11), 118701 (2004)
41. D. Halliday, R. Resnick, J. Walker, *Fundamentals of Physics* (Wiley, New York, 1993)
42. P. Grassberger, Some more exact enumeration results for 1d cellular automata. J. Phys. A: Math. Gen. **20**(12), 4039–4046 (1987)

Chapter 6
Information Dynamics in Networks and Phase Transitions

> "Understanding the ways in which information spreads in networks is one of the most important open problems in science." Mitchell, 2009 [1].

As outlined in Sect. 2.4, the topology of networks has attracted much recent attention, however their time-series dynamics remains relatively poorly understood. In particular, the importance of quantitatively establishing the nature of *distributed computation* in networks is widely acknowledged, e.g. Mitchell [2] states "the main challenge is understanding the dynamics of the propagation of information ... in networks, and how these networks process such information."

As compared to other attempts to describe information manipulation in networks (e.g. [3–8], discussed in Sect. 2.4), our perspective of how information is acted upon in intrinsic *distributed computation* in these network types is an important one. It is underlined by the comments of Mitchell above on information dynamics in networks, and by the general importance attributed to information processing in biological networks [9–11]. Crucially, our framework is unique in quantitatively aligning with popular understanding of information storage, transfer and modification (as per Chaps. 3, 4, and 5).

In this chapter, we examine the information dynamics of networks using the framework introduced in Chaps. 3–5. This examination takes the perspective of the distributed computation undertaken by the nodes in the intrinsic computation of their attractor. We study two important models of time-series dynamics on networks: random Boolean networks (RBNs), and cascading failures in power grids. RBNs were introduced in Sect. 2.4.1, while a model for cascading failures will be introduced in Sect. 6.2.1. Both network models undergo a *phase transition* between ordered and chaotic dynamics as a key parameter is altered (average connectivity for RBNs, network tolerance for the failures model).

As described in Sect. 2.4, it has been hypothesised that networks close to the critical state possess a maximal information transfer capability (e.g. [6]). This is generalised in the *edge of chaos* hypothesis [12]: that systems exhibiting critical dynamics in the vicinity of a phase transition *maximise* their computational properties (see [11, 13] regarding RBNs in particular). More specifically, Langton [12]

J. T. Lizier, *The Local Information Dynamics of Distributed Computation in Complex Systems*, Springer Theses, DOI: 10.1007/978-3-642-32952-4_6,
© Springer-Verlag Berlin Heidelberg 2013

suggests that *intermediate levels* of information transfer and storage give rise to complex computation in critical dynamics, with too much of either decaying the computational capability. This is at odds with suggestions of the maximisation of information transfer in this regime, e.g. [3, 6, 14, 15].

Such conjectures can be investigated with our framework now in place. Our experiments will quantify the *average* information dynamics in summarising the *ensemble* properties of the networks in regard to the underlying phase transitions in their behaviour. It is simple to foresee the average active information storage and apparent transfer entropy (TE) being zero in the extreme ordered regime (with fast freezing at point attractors) and in the extreme chaotic regime (where the high level of interactions overwhelm information storage and obscure the apparent contribution of each information source). It seems reasonable that both would be maximised, on average, in the interim near the critical region, where the dynamics support long correlations across space and time. On the other hand, we predict that the complete TE (which captures information contributions due to interactions between multiple sources, see Sects. 4.1.3 and 4.2.4.2) will continue to increase with the connectivity into the chaotic regime. Indeed, we observed that relatively high values of the apparent TE were associated with the capacity for coherent local information transfer structures (i.e. gliders in CAs). We hypothesised in Sect. 4.2.4.2 that an increasing of the complete TE in the chaotic regime indicated a higher level of interactions in conjunction with the loss of this coherence.

Important caveats are provided though by criticisms of the edge of chaos hypothesis, e.g. see [16, 17]. In examining *average* computational properties as a function of RBN parameters, we emphasise that there is in general a very large range of network realisations and consequently of behaviours possible for each parameter set [18].[1] The local information dynamics of computation will provide much more detailed insights *for a given* RBN (as we have seen for CAs) than averages over nodes, networks and network sets discussed here. That being said, these ensemble averages can provide important insights into the computational properties as a function of RBN parameters. We must simply remember that the average results are akin to likelihoods rather than certainties, albeit likelihoods that are much stronger in the limit of infinite system size where the phase transition becomes discontinuous [7].[2] In contrast to some of the criticised studies, we emphasise that the models investigated here exhibit well-established transitions with respect to a single order-chaos parameter. We also emphasise that in quantifying the elements of computation for any RBN, application of our framework will uphold criticisms in [16] in showing that computation does not only occur at an edge of chaos, but occurs to *some extent* at *all phases* of behaviour.

In this chapter, we report the results of our analyses on RBNs in Sect. 6.1 and the model of cascading failures in Sect. 6.2.[3] We demonstrate in both network models

[1] This is commonly observed in such ensemble studies [17, 19].

[2] See also the see sharpening of phase transition with system size for RBNs in [7, 20] and for another network type in [3].

[3] Our information dynamics analysis of RBNs was first reported in [21], and the analysis of cascading failures first reported in [22].

that information storage and coherent (or single-source) transfer are each maximised in the vicinity of the phase transition between ordered and chaotic dynamics. Importantly, we demonstrate a shift from the dynamics being dominated by information storage in the ordered regime, to a balance of information storage and transfer around the critical point, and a further shift to the dominance of interaction-based information transfer in the chaotic regime. These findings align with much conjecture regarding computational properties of networks and other systems undergoing phase transitions [3, 6, 7, 12]. Near the critical point we observe maximum capability for coherent computation, inferred by maximisations of active information storage, apparent TE, and relatively few but high-impact non-trivial information modification events. Beyond this in the chaotic regime however, the interaction between the nodes begins to dominate (inferred by increases in complete TE but a reduction in apparent TE). This erodes the capacity for coherent computation in this regime. Furthermore, in using our two complementary measures of information transfer (apparent and complete TE), we are able to resolve the paradox that a too large an amount of damage spreading or cascading failures actually means a smaller amount of coherent or single-source information transfer. Finally, we also reveal interesting relationships between network structure and dynamics, in particular the role of hubs in increasing information transfer in networks.

6.1 Phase Transitions in Random Boolean Networks

In this section, we seek to measure the *average* information dynamics during the phase transition in RBNs. This phase transition occurs as a function of average in-degree or connectivity $\overline{K}$, with the networks exhibiting ordered behaviour at low $\overline{K}$, and chaotic behaviour at large $\overline{K}$. We will describe our experimental approach in Sect. 6.1.1, then describe our results in Sect. 6.1.2.

6.1.1 Experimental Details

For the RBNs simulated here, we use $N = 250$, Poissonian distributed in-degree for each node based on average in-degree $\overline{K}$, $p = 0.5$ (no bias in rules), and classical RBNs (CRBNs) with synchronous updating (see Sect. 2.4.1). Also, we do not bias the network structure, allowing comparison with the majority of existing RBN publications. The RBNs are modelled using enhancements to Gershenson's RBNLab software [23].

The phase transition in these RBNs is traditionally quantified using a measure of sensitivity to initial conditions, or damage spreading. Following [18], we take a random initial state A of the network, invert the value of a single node to produce state B, then run both A and B for many time steps (enough to reach an attractor is most appropriate). We then use the Hamming distance:

$$D(A, B) = \frac{1}{N} \sum_{i=1}^{N} |a_i - b_i|, \tag{6.1}$$

between A and B at their initial and final states to obtain a convergence/divergence parameter δ:

$$\delta = D(A, B)_{t \to \infty} - D(A, B)_{t=0}. \tag{6.2}$$

(Note $D(A, B)_{t=0} = 1/N$). Finding $\delta < 0$, implies the convergence of similar initial states, while $\delta > 0$ implies their divergence. For fixed p, the critical value of $\overline{K}$ between the ordered and chaotic phases is [24]:

$$K_c = \frac{1}{2p(1-p)}. \tag{6.3}$$

For $p = 0.5$, we have $K_c = 2.0$. The standard deviation of δ peaks slightly inside the chaotic regime for finite-sized networks, indicating the widest diversity of networks for those parameters [20].

We take an ensemble approach to measuring the average information dynamics as a function of $\overline{K}$. To measure the active information storage as a function of $\overline{K}$ for example (denoted as $A_X(k, \overline{K})$), we:

1. measure $A_X(k)$ for each node in a given RBN generated with $\overline{K}$;
2. then average these over each node in the RBN to get $\langle A_X(k) \rangle$ for the network;
3. then average these network averages over many networks generated for each $\overline{K}$ (at least 250) to determine the average value $A_X(k, \overline{K})$ as a function of $\overline{K}$.

We measure the average entropy and entropy rate as a function $\overline{K}$ of in this way also. Similarly, the average apparent and complete transfer entropies are measured for (at least 50) sample pairs of *causally linked* nodes,[4] averaged once to obtain network averages, and again over many networks to obtain averages as a function of $\overline{K}$. A hybrid local-average approach is taken for the separable information; the average $S(k, \overline{K})$ is computed in a similar manner to the other measures, however we also record the balance between its positive and negative local values (trivial and non-trivial information modifications respectively) $S_X^+(k, \overline{K})$ and $S_X^-(k, \overline{K})$ in contributing to the average (see Eq. (5.8)).

We note that larger networks have more sharply-defined order-chaos phase transitions and less variation in ensemble properties as a function of $\overline{K}$ [6, 7]. Indeed, the edge of chaos hypothesis focuses on computation in the limit of infinitely-sized networks, where the phase transition exhibits *discontinuous* changes in system properties with respect to continuous variation of parameters (see Sect. 2.4.1). To assist

[4] This is unlike the mutual information measurements by Ribeiro et al. [7] and Solé and Valverde [3] for *random* node pairs (regardless of whether they are directly causally linked).

our finite-size networks to approximate this phase transition, we avoid running the RBN for too many time steps. This is because the computation is completed once the network reaches a periodic or fixed attractor (see Sect. 4.2.3), which is inevitable for finite-sized RBNs. For each simulation from an initial randomised state, we ignore a short initial transient of 30 steps to allow the network to settle into the main phase of the computation, then allow evolution over 400 time steps. Importantly, since the nodes in each RBN are heterogeneous agents, the PDFs for each measure must be computed for each node individually rather than combining observations across all nodes.[5] In order to properly sample the dynamics of each node in each RBN and generate enough data for the information-theoretic calculations, many repeat runs from random initial states are required for each network (at least 4480 are used). For these calculations, one should use as large a history length k as facilitated by the number of observations in order to separate information storage and transfer (e.g. see Sect. 3.3.1). Here we find $k \approx 13$ provides reasonable convergence for a reasonable number of repeat runs.

6.1.2 Results and Discussion

6.1.2.1 Information Storage and Transfer Components

Figure 6.1 shows that the average single node entropy $H_X(\overline{K})$ simply increases as a function of $\overline{K}$, as expected since the level of activity in the network is increasing with this parameter. More importantly, Fig. 6.1 also plots the average active information storage $A_X(k = 14, \overline{K})$ and entropy rate $H_{\mu X}(k = 14, \overline{K})$. This figure shows that the active information storage rises then reaches a maximum near to the critical phase ($\overline{K} = 2$) before falling away, while the entropy rate only begins to rise near the critical phase then continues to rise and approach the entropy in the chaotic phase. Since the entropy is the sum of the active information storage and entropy rate (Eq. 3.21), which is equal to collective information transfer in this deterministic system (see Sect. 4.1.4), we can now begin to describe the phase transition in terms of computation:

- **the ordered phase is dominated by information storage** (information contained in the past of the node about its next state);
- **the chaotic phase is dominated by information transfer** (information from incoming links about the next state which was not contained in the node's past);
- while there appears to be something of **a balance between the two near the critical phase**.

[5] In contrast, this could be done for the homogeneous agents in CAs in the preceding chapters.

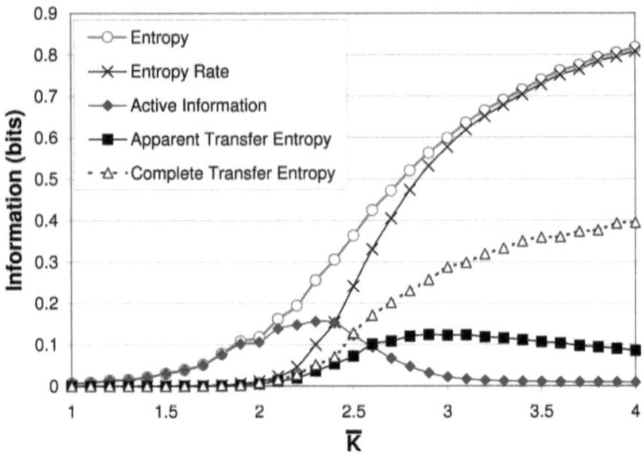

Fig. 6.1 Average information dynamics versus average connectivity $\overline{K}$ for networks of size $N = 250$. Plotted here are the average entropy $H_X(\overline{K})$, entropy rate $H_{\mu X}(k = 14, \overline{K})$, active information $A_X(k = 14, \overline{K})$, apparent transfer entropy $T_{Y \to X}(k = 14, \overline{K})$ and complete transfer entropy $T^c_{Y \to X}(k = 13, \overline{K})$. The information required to predict the next state of each node is dominated by information storage at low $\overline{K}$ and by information transfer at higher $\overline{K}$ (first by coherent then interaction effects). Error bars (omitted) are on the scale of the data points for all plots (NB: This figure was first published in [21].)

6.1.2.2 Coherent and Interaction-Based Transfer

We then examine the constituency of the information contributed from incoming links, the total of which is the entropy rate (see e.g. Eq. (4.47)). Figure 6.1 also plots the average *apparent* TE $T_{Y \to X}(k = 14, \overline{K})$ for each link, demonstrating that this quantity also rises to a maximum value close to the critical phase, then falls away. In contrast, Fig. 6.1 additionally plots the average *complete* TE $T^c_{Y \to X}(k = 13, \overline{K})$ for each link. This measure also begins to rise close to the critical phase but continues to increase into the chaotic phase.

We see therefore that in the first stage of the shift toward the dominance of information transfer, single sources can be observed to have a significant influence on the destination (in the context of the destination's history) without considering the effect of the other causal sources (i.e. $T_{Y \to X}(k = 14, \overline{K})$ is relatively high). **In the critical regime, there is maximum potential for single-source influences to be propagated as *coherent* information transfer structures**. However, as the activity level in the RBNs continues to rise with the average connectivity $\overline{K}$, the apparent effect of each source is swamped by the activity of the other causal sources. That is: an observer cannot discern the effect of a single source, leading $T_{Y \to X}(k = 14, \overline{K})$ to fall away. Considering then the increase in $T^c_{Y \to X}(k = 13, \overline{K})$ (which in accounting for the other sources also detects interaction-based transfer), we see that **the level of interaction is increasing with the connectivity of the network**. Also, **in the chaotic regime the influence of any one information source can only be properly identified by taking all of the other sources into account**.

These complementary measures of information transfer provide different but useful insights, and give impetus to our hypothesis in Sect. 4.2.4.2 regarding the relative values of the apparent and complete components of information transfer in order-chaos phase transitions. Indeed, we can provide quantitative evidence to the conflicting conjecture around whether information transfer is found at an intermediate [12] or maximum level [3] at criticality. For RBNs, **transfer is maximised close to criticality where one measures the apparent influence of a source in isolation,but equally it is at an intermediate level where the measurement considers transfer due to interaction with other causal information sources also**. If these findings apply to such phase transitions in general, then both sources of conjecture appear to be well-founded, being resolved in these two different methods of measuring information transfer.

6.1.2.3 Location of Maximisations of Computational Capabilities

Next, we compare these maximisations to the location of the phase transition as measured using the standard deviation of the convergence/divergence parameter δ (from Eq. 6.2).[6] In Fig. 6.2 we see that the **information storage peaks slightly within the ordered phase from the critical region, while the coherent information transfer peaks slightly within the chaotic phase**. Importantly, it is the apparent TE that peaks here (indicating the capability for coherent information transfer), as distinct from the complete TE which continues to increase into the chaotic phase. As per footnote 6, we expect the relative positions of these maximisations to be maintained around the critical phase as $N \rightarrow \infty$, with both likely to become closer to the critical point in this limit (as for the measure of correlation by Ribeiro et al. [7]). The relative positions of the maximisations are quite interesting, because they align with existing conjecture on the nature of computation around phase transitions which typically associates information storage with the ordered phase and information transfer with the chaotic phase (e.g. [12]).

Additionally, we note that it is not unusual for different measures of complexity or structure to be maximised at different parameter settings, because they are measuring different properties of the computation in the system and are not trivially related [25]. We need measures of both information storage and transfer to understand the distributed computation taking place in the system.

---

[6] δ was confirmed to change sign close to $\overline{K} = 2$ here (as per [20]), with a subsequent slow increase after $\overline{K} = 2$ (known to be a finite-N effect). The standard deviation of δ is maximised during this increase in the chaotic regime [20]. Certain other measures suggested to indicate the critical phase are known to be shifted into the chaotic regime for finite-N, e.g. [7]. Given impetus as an indicator of the critical phase by the related measure of Rämö et al. [6], we use the standard deviation of δ as guide to the relative regions of dynamics in finite-N networks.

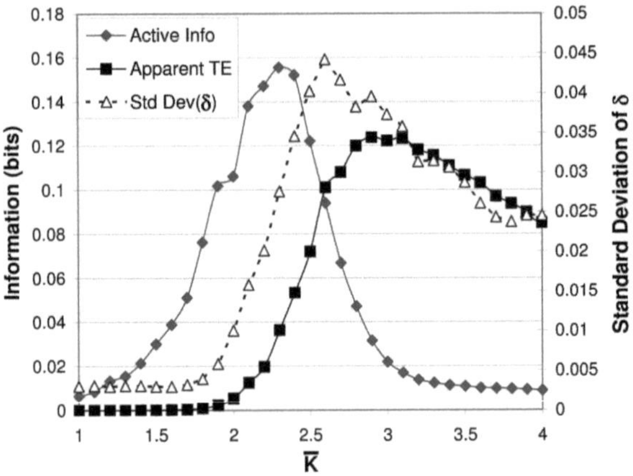

Fig. 6.2 Maximisations in active information $A_X(k = 14, \overline{K})$ and apparent transfer entropy $T_{Y \to X}(k = 14, \overline{K})$ as a function of average connectivity $\overline{K}$ for $N = 250$, shown with respect to the standard deviation of the convergence/divergence parameter δ. This indicates that information storage peaks just on the *ordered* side of the phase transition, while (coherent) information transfer peaks just on the *chaotic* side of the phase transition (NB: This figure was first published in [21].)

6.1.2.4 Information Modification and Coherent Computation

We have suggested above that relatively large values of the apparent TE can be interpreted as a capacity for the coherence of information transfer. Further insight into the coherent information structure in the computation in the RBN is provided by the separable information $S_X(k, \overline{K})$. Figure 6.3 shows that $S_X(k, \overline{K})$ is maximised for approximately the same values of $\overline{K}$ as the apparent TE (though it is slightly more spread out). This can be explained with reference to its positive and negative components, $S_X^+(k, \overline{K})$ and $S_X^-(k, \overline{K})$. We see from Fig. 6.3 that the early rise in the separable information is driven by $S_X^+(k, \overline{K})$ (trivial information modifications), with a peak occurring before $S_X^-(k, \overline{K})$ (non-trivial information modification events) rises and consequently reduces the total. As the connectivity $\overline{K}$ is further increased, $S_X^+(k, \overline{K})$ begins to fall whereas $S_X^-(k, \overline{K})$ continues to rise. **Near the critical phase**, at the peak of the separable information, note that **there is in fact a relatively low incidence of non-trivial information modification events** (i.e. $S_X^-(k, \overline{K})$ is low). This is interesting because of the importance placed on these events in computation, e.g. they are manifested as particle collisions in CAs Sect. 5.2.1. **It appears that if the amount of non-trivial information modification events or information collisions is too large, the capacity of the system for complex computation is reduced. It is likely that this is due to a large amount of "information collisions" eroding the coherent nature of the information storage and transfer within the system, disturbing the computation and reducing their own impact**. A maximisation of separable information should perhaps be interpreted as maximising the bandwidth

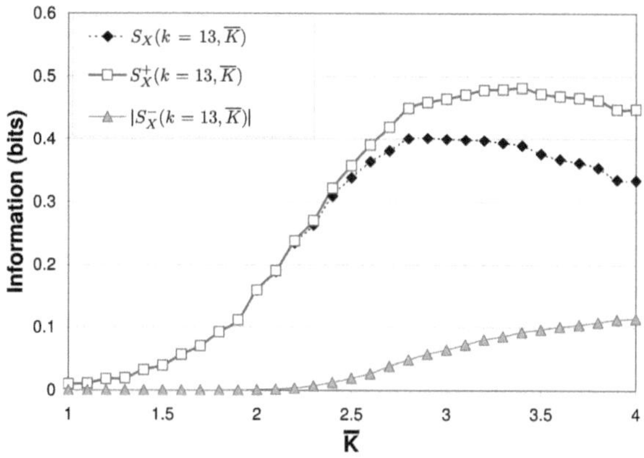

Fig. 6.3 Separable information $S_X(k = 13, \overline{K})$ and its positive and negative components, $S_X^+(k = 13, \overline{K})$ and $S_X^-(k = 13, \overline{K})$ respectively, versus average connectivity $\overline{K}$ for $N = 250$. Trivial information modification (high $S_X^+(k)$) dominates the dynamics at low $\overline{K}$, while the amount of non-trivial information modification rises with $\overline{K}$ (NB: This figure was first published in [21].)

for coherent information storage and transfer, while allowing a smaller number of *high-impact* non-trivial information modification events in the coherent computation.

Finally, we note that all of the information dynamics described here experience maximum standard deviation in the vicinity of the critical region (not shown). This indicates maximal diversity in the information dynamics throughout the RBNs in this regime, as observed for other measures (e.g. [20]).

6.2 Cascading Failures in Power Grids

Modern energy, communication, transportation and financial networks have evolved in various complex ways to satisfy the demands of their end-users under various resource constraints, but of some concern is that they are all subject to *cascading failure events* [26–28]: local failures that trigger avalanche mechanisms with large effects over the whole network. Our focus is energy networks, where usage has increased faster than investment in infrastructure (much of which is reaching the end of its useful life) and the grid has become critically loaded. As such, small failures can lead to cascading catastrophic blackouts and outages are occurring more frequently (see examples in [26]) and with more adverse impact. These can disrupt important services and cost millions of dollars. It is important to understand these events so that they may be avoided.

Much conjecture regarding information dynamics in networks surrounds these cascading failures and the related phenomena of damage spreading/perturbation avalanches. For example, [10] discusses the phase transition in RBNs in terms of damage spreading: in the ordered phase, single-node perturbations usually die out, in the chaotic phase they tend to propagate through the entire network, whilst in the critical regime there is maximal uncertainty in their propagation. Indeed, in [6] information propagation is measured as the uncertainty in avalanche size. Also, discussions of information transfer in cells typically focuses on propagating cascades of "signal" elements [11, 29]. Similar too are waves of directional change in schooling fish, referred to as "information cascades" in [30], and waves of change in stomatal aperture in plants [31]. The analogy to gliders in CAs for all these coherent cascading effects is clear, and indeed observed in [31]. Memory has also previously been studied in cascade-style systems in [32] using correlation coefficients of inter-event times. Without proper quantification of the information dynamics of these phenomena though, it remains unclear whether for example, information transfer should be directly identified with damage spreading itself, or perhaps with the uncertainty in the extent of damage spreading. Indeed, Mitchell [1] states that "the phenomena of cascading failures emphasises the need to understand information spreading and how it is affected by network structure."

We introduce a model used for studying cascading failures in networks in Sect. 6.2.1, which exhibits a phase transition between ordered and chaotic responses to single-point failures with respect to failure tolerance in the network. We then describe in Sect. 6.2.2 how we will study computation during cascading failures using this model. Subsequently we present the results of our analysis in Sect. 6.2.3, which echo our findings in RBNs of maximisation of information storage and (coherent) transfer properties near the critical phase.

6.2.1 Cascading Failures Model

Cascading failures are studied here using the model described in [26], and subsequently used e.g. in [33, 34]. We focus on the use of this model to study energy networks (as per [33]), however note that it is also applicable to communication or transport networks, e.g. the Internet. The network is constructed as a weighted, undirected graph of N nodes (representing substations for our purposes) and K edges (representing the transmission lines). The model describes the weights of each edge and loads of each node, and how these interdependently evolve after the breakdown of a node.

Each edge has an **efficiency** $e_{ij}(n)$ (akin to relative capacity in the energy networks analogy) that changes with time n. If there is an edge from node i to j then $e_{ij}(n) \in (0, 1]$ and is initialised to $e_{ij}(0) = 1$ (if there is no edge $e_{ij}(n) = 0$). Edge efficiency is the inverse of edge weight, while the efficiency $\epsilon_{ij}(n)$ of the most efficient *path* from i to j is the inverse of the shortest path length [8].

Each node i has a **load** $L_i(n)$, being the total number of most efficient paths passing through it at time n (i.e. the load is the *betweenness centrality* of the node) [35]. Each node is also assigned a **capacity** C_i, being the maximum load it can handle without performance degradation. The capacity is assumed to be proportional to the initial load of the node: $C_i = \alpha L_i(0)$ [27], with $\alpha \geq 1$ being the *fixed* network tolerance.

The edge efficiencies become sub-optimal if[7] *either* end-point node is operating above capacity:

$$e_{ij}(n+1) = \begin{cases} e_{ij}(0) \min\left(\frac{C_i}{L_i(n)}, \frac{C_j}{L_j(n)}\right) & \text{if } L_i(n) > C_i \text{ or } L_j(n) > C_j, \\ e_{ij}(0) & \text{otherwise.} \end{cases} \tag{6.4}$$

Changes induced in edge efficiencies by excess loads cause changes in most efficient paths and therefore the distribution of loads at the next time step. While the initial state of the network is stable (since $\alpha \geq 1.0$), the removal of a node (simulating the breakdown of a substation) triggers a dynamical process where the network loads are redistributed. This process can cause other nodes to overload, shunting their loads onwards to other nodes who then overload etc., thereby stimulating a cascading failure in time. Effectively, the network is computing its new stable state (or attractor) during these events.

The performance of the network as a whole during these events is tracked using the **average efficiency** between all node pairs, $E(n) = \langle \epsilon_{ij}(n) \rangle$. Typically, for large α the network efficiency is relatively unaffected by node removal, however as α becomes closer to 1 the lower tolerance means that node removal can have dramatic cascading effects (see Fig. 2.4 in [26]). Indeed the average network efficiency after node removals falls very quickly with α as $\alpha \to 1$, moving from a stable (efficient) state to an unstable one. We note that altering α here determines the phase of the network's dynamic response in a similar way to the average connectivity in RBNs: for large α the response is in a somewhat ordered phase, for α close to 1 it is chaotic, and in a critical phase between these. In the next section, we describe how the information dynamics of this phase change will be measured.

6.2.2 Measuring Information Dynamics in Cascading Failures

In our preceding study of RBNs, we compared damage spreading and information dynamics properties of RBNs as a function of $\overline{K}$, but the information dynamics properties were measured for the *average* activity of the networks. Here we seek to examine the information dynamics specifically *during* avalanches or cascading failures as a function of tolerance α. From that preceding study, we could conjecture information transfer measured by complete TE to be related to avalanche size, since both increase into the chaotic regime. Similarly, information transfer measured by apparent TE could be more closely related to uncertainty in avalanche size, since both are maximised near the critical phase. More important than these coincidences however is the intuition that in the chaotic phase where avalanches are typically large and overlapping, the damage spreading effect of one node on another will manifest via interactions with other sources. This will only be captured by the complete TE. On the other hand, near critical point there is maximum differentiation of the effect

[7] We have slightly altered the definition in [26], since this definition was ambiguous.

of one node on another, which will be captured by the apparent TE; indeed it should thus be maximised near this phase transition where coherent cascades (in analogy to gliders in CAs) have clear causes.

In this context, going beyond studying damage spreading in a generic model such as RBNs we study the information dynamics of cascading failures in power grids, since there is significant social and economic impetus for focusing on this application. Here, we use the model presented in the previous section with the topology of the electrical power grid of the western United States [36] ($N = 4941$ and $K = 6594$). For network tolerances $1.0 \leq \alpha \leq 1.3$, we simulate the breakdown of substations by removing 149 randomly selected nodes (one at a time) and then measure the resulting time series of node loads and edge efficiencies until the network reaches a (possibly periodical) stable attractor state.

For each tolerance α, the active information storage A_i is measured from the loads for every node i, and the transfer entropies $T_{i \to j}$ and $T_{j \to i}$ are measured in each direction for each causal link ij between their time series of loads. We use a history length $k = 3$, which is limited by the amount of simulated data that we could reasonably produce but (qualitatively) appears sufficient for the dynamics here. PDFs are estimated for each individual node and linked pair from the set of time series of observations of *loads* obtained from all the separate node knock-outs, using kernel estimation (see Sect. 2.2.3) with a kernel width of 0.5 standard deviations of each variable.

6.2.3 Results and Discussion

6.2.3.1 Maximisations of Information Dynamics Near the Phase Transition

First, we examine the average apparent TE (across all links, $T = \langle T_{i \to j} \rangle$) and average active information storage (across all nodes, $A = \langle A_i \rangle$) in the network as a function of tolerance α. Figure 6.4 shows that with large α **in the ordered phase, information storage is the dominant function**. As the tolerance is decreased, **both storage and transfer reflect the phase transition in average network efficiency**, **being maximised in the vicinity of the critical state**. To some extent one would expected that information transfer is (at least initially) increased as the efficiency α falls, since cascades of larger and larger sizes are occurring. At first glance then it may be surprising that maximisations in these measures occur *before* the system is pushed into a chaotic state at $\alpha = 1.0$. However, these results are in alignment with those from RBNs in Sect. 6.1.2 and our hypotheses in Sect. 6.2.2. As per the shift towards the chaotic phase in RBNs, the increased activity and interactions in the network here as $\alpha \to 1.0$ begins to obscure the individual contribution of any source node to a given destination, so **measurement of single-source information transfer falls in the chaotic phase**. Also similar to the results from RBNs is that the maximisations in active information storage and apparent TE are closer to the ordered and chaotic phases respectively.

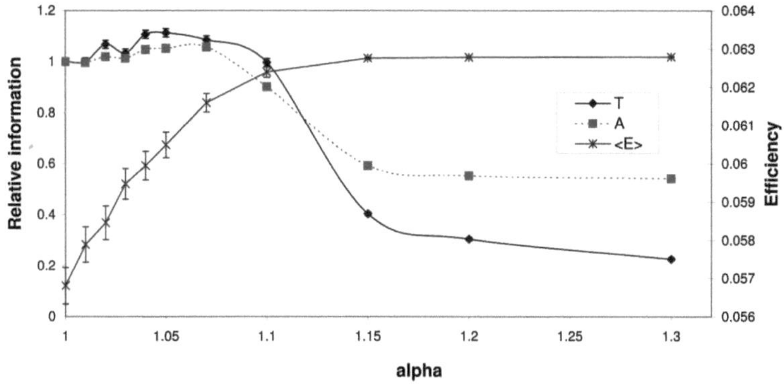

Fig. 6.4 Information dynamics versus network tolerance α, plotted relative to their values for $\alpha = 1.00$. Transfer entropy T is averaged over all time steps for each link, then averaged over all links, with $T(\alpha = 1.00) = 0.0155$ bits. Similarly, active information storage A is averaged over all nodes, with $A(\alpha = 1.00) = 1.10$ bits. Also plotted is the final network efficiency $\langle E \rangle$ averaged over each simulated node removal. Error bars for T and A are on the scale of the data points (NB: This figure was first published in [22].)

Another particularly interesting result is that as α is reduced **the information dynamics change more rapidly than network efficiency** (see $1.1 \leq \alpha \leq 1.15$ in Fig. 6.4) and *could be a useful early indicator of critical loading*. To clarify: this result does not imply the information dynamics would provide an earlier detection during any particular cascade. It does imply that, in measuring the response to failures in the network they may be more useful in detecting that the network is actually critically loaded and prone to large cascading failures.

We also plot the *accumulated* local information transfer and storage in Fig. 6.5 (i.e. the sum total of each throughout the time of the cascade rather than the average over time). These values continue to rise towards the chaotic regime, in contrast to the decrease in *average* information transfer and storage at each time step in the cascade. The *total* information transferred and stored during the cascade increases simply as a legacy of the longer cascade lengths.

6.2.3.2 Local Information Transfer at Each Node: Relation to Topological Features

We then examine the information transfer values *locally* at each link, $T_{i \to j}$, checking for any relation between the topological features and the information dynamics they give rise to. Indeed, similar questions have been considered with the TE in the context of cortical networks, with the authors of [37] surprised to find large differences between different nodes (i.e. a non-"democratic" distribution of TE), while large variations are observed over time for hubs in [38]. Interestingly, this approach brings

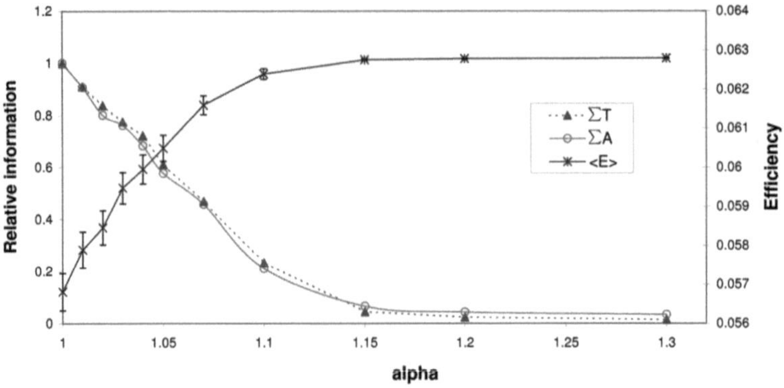

Fig. 6.5 Accumulated information dynamics versus network tolerance α, plotted relative to their values for $\alpha = 1.00$. Transfer entropy T and active information storage A are sum over all time steps for each link/node for each node knock out, $\sum t_{i \to j}(\alpha = 1.00) = 7.69 \times 10^5$ bits, $\sum a_i(\alpha = 1.00) = 2.04 \times 10^7$ bits. Also plotted is the final network efficiency $\langle E \rangle$ averaged over each simulated node removal

together our local view of information dynamics with the trend towards using a local view of topological measures (e.g. see [39, 40]).

Here, Fig. 6.6 shows a **strong correlation of the transfer entropy to *source* node degree** (with the TE averaged over all applicable links). Correlations are large ($r \geq 0.80$) and significant ($p < 0.01$) for α in the vicinity of the phase transition, becoming somewhat weaker in the stable region. Similar results are found for active information storage with node degree. While we know that failures in nodes with larger degree typically lead to larger cascades [26], this does not explain why nodes with larger degree transfer (and store) more information since as we have seen above that larger cascades (in the chaotic region) do not equate to larger information transfer. A key factor in explaining these effects is that the links are bidirectional here, so "degree" is equivalent to both "in-degree" and "out-degree". The more neighbours a node has, the more locations it has to store and retrieve information, and the more sources it has from which to obtain information that it can then transfer. From a similar perspective, the more neighbours a node has, the more sources of input it has and the more *diversity* in its activity: this intrinsic variation in a node is a dominant factor in how much information it can both store and transfer. Indeed, this is why nodes with degree one have zero information in this model: with zero betweenness centrality their load never changes, so there is no diversity that they can transfer and no influence on them to measure. Another factor may be that the more neighbours a source node has, the more likely that its information content is novel for the destination, allowing larger information transfer. Indeed, this aligns with the view that hubs increase information exchange between "otherwise distant or disconnected nodes" [38]. The key result here is that **hubs are the most informative sources**. Importantly, this is not the trivial statement that they transfer more information in total because they have more outgoing links to transfer information on; the statement is that more information is transferred down *each* link sourced from the hub. This

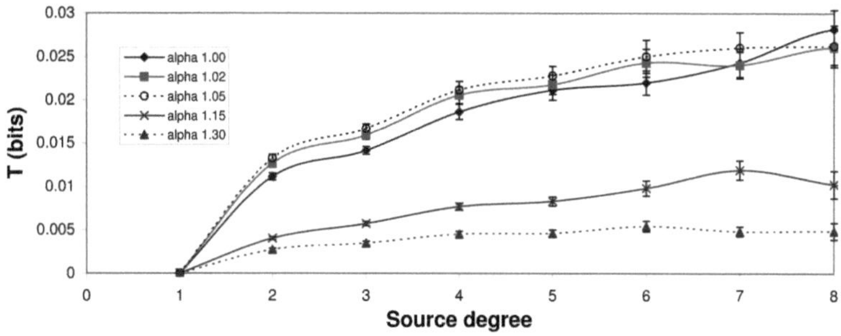

Fig. 6.6 Average transfer entropy T versus degree of the source node for various α. The small number of nodes with degree larger than eight introduces large errors beyond this point (NB: This figure was first published in [22].)

may indeed explain the (application-specific) observation that scale-free networks spread information "better" than random networks [41].

We note that it is quite plausible that in many real-world networks the desirability of connecting to another node is proportional to the information that can be transferred from that node (since there is a higher potential reward in information for the cost of the link). Where this is the case, then this correlation of information transfer to source degree could be a generic, functional driver for the important process of *preferential attachment* [42, 43] (see Sect. 2.4) in undirected networks. The undirected nature of the links is important in providing a positive feedback component to the process, because in drawing an input from a hub the new node also adds a new input to that hub, thereby increasing the average information transfer from the hub even further.

On the other side of the links, **there is no corresponding correlation of the transfer entropy to the degree of the destination node**. Certainly, the increased diversity in the destination provides greater scope for information transfer to it, but the more difficult it becomes to identify the coherent effect of the source[8]: these effects seems to counter each other. We also measure small but significant (at $p < 0.001$) **correlations between the initial load of a node (i.e. its betweenness centrality) and the information it stores and transfers to other nodes** during the cascade (e.g. for $\alpha = 1.05$ these correlation coefficients are 0.24 and 0.13 respectively). In a similar fashion to degree, as the betweenness centrality of a node increases so too does its propensity to be influenced by other nodes, and therefore the diversity of its activity increases.

6.2.3.3 Local Information Transfer in Time: How the Cascade Unfolds

Finally, we examine how the local information dynamics evolve in time as the cascade unfolds. Figure 6.7 shows that for $\alpha = 1.05$ the local information transfer values

8 It could also be said that there is a large amount of assortative noise between the source and destination [3, 44].

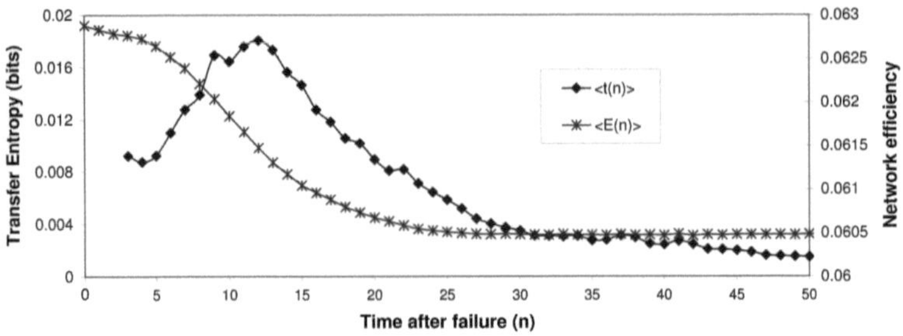

Fig. 6.7 Local transfer entropies in time (averaged across all links for the given time step n: $\langle t_{i\to j}(n)\rangle_{ij}$) and network efficiency $E(n)$ versus time after initial node removal. Both measures are averaged across all simulated node knock outs with $\alpha = 1.05$ (NB: This figure was first published in [22].)

exhibit a strong peak around 12 time steps after the cascading failure (similar plots are observed for other values of α). This peak lags the initial failure event but coincides with the steepest drop in network efficiency. The cascade takes some time to build up in size, but causes the steepest change in network efficiency and largest information transfer when it is spreading at its most rapid. **The relation between the *change* in network efficiency and local information transfer (in time) is quite strong**: they have a correlation of −0.94. The local information transfer could therefore be a useful *application-independent* indicator of the spread of the cascade, in particular its peak as well as its subsequent passing. Indeed, were the local values to be examined in space (i.e. on particular links) as well as time, they could be used as indicators of the direction of spread of the cascade. Interestingly though, the local values in space and time do not exhibit a strong correlation to local loads—it appears the utility of the local information transfer in time discussed above is an self-organised emergent effect which can be seen only at the macroscopic level.

6.3 Summary

We have described results which quantify the fundamental nature of computation around the critical phase in two models of network dynamics. The dynamics here are dominated by information storage in the ordered phase, with the level of information storage increasing with activity in the network (facilitated in RBNs by increasing connectivity). The increased activity also gives rise to an increasing level of information transferred from linked nodes. These two operations of universal computation appear to be in balance around the critical point. After this, information transfer continues to increase for a period, reducing the capacity for information storage. In RBNs we observed that near the critical point there is a large amount of trivial information modifications, providing the capability for coherent information transfer and storage to flourish and indeed maximise, and allowing the small number of non-trivial information modifications to have a large impact on the coherent computation. As activity continues to increase, the coherent transfer from any single node observed in isolation initially appears strong, with the apparent TE peaking slightly into the chaotic regime. With further increases however, the interaction between the nodes begins to dominate, being manifested in large multi-source transfer. This erodes the capacity for coherent computation.

We emphasise the key insight gained here on conflicting conjectures regarding whether information transfer peaks near criticality or continues into the chaotic regime, or equally whether more variable or larger cascade size should be identified with information transfer. We resolved this in revealing the subtle relationship between the two different perspectives of information transfer (via apparent and complete TE). Apparent TE measures transfer attributable to a single node observed in isolation and therefore decreases in the chaotic regime. In contrast, the manner in which complete TE conditions on other nodes captures multi-source interactions also and therefore increases in the chaotic regime.

Furthermore, we have revealed interesting relationships between underlying topology (average connectivity for RBNs, and node degree in the cascading failures model) and the local information dynamics they give rise to. We also demonstrated a strong relationship between the spreading of cascades (in an application-*dependent* level) and information transfer across the network (in an application-*independent* manner): this is important since a key aim of the application of information theory to complex systems is to provide application-independent explanations of the dynamics.

The new understanding of the information dynamics in RBNs is important given their role in modelling computation in biological networks [10, 11, 13, 29, 45]. Our findings surrounding RBNs near the critical phase are of particular interest, because there is evidence that the Gene Regulatory Networks they model operate in this critical regime [46]. The implication here is that GRNs have evolved to a form facilitating maximum coherent computational capability.

Our application to cascading failures is important in that the results are applicable to the many related phenomena, including damage spreading or perturbation avalanches [6], and coherent signal cascade phenomena [11, 29, 30]. Indeed the maximisation of apparent TE while the cascading effect remains coherent (before the chaotic phase) underlines the analogy of this phenomena to gliders in CAs (e.g. [31]). Furthermore, the fact that information dynamics initially changed faster than network efficiency as the tolerance α drops suggests that information dynamics may indeed be more useful than traditional measures in detecting that the network is critically loaded.

We also emphasise that this study of RBNs and cascading failures represents the first explorations of order-chaos phase transitions using this framework for information dynamics. Our findings here align with much conjecture regarding computational properties of not only networks [6, 7] but other systems undergoing phase transitions [12, 14]. As such, the results here are likely to be applicable to similar phase transitions in other systems, i.e. those simple transitions controlled by a single order-chaos parameter (as approximated in our finite-sized systems here). These may include flocking behaviour [30, 47] and ant foraging [14]. As discussed in the next chapter though, the results are not likely to be universal to more complicated transitions.

While our results have certainly provided quantitative answers for much conjecture about the fundamental nature of the computational properties of networks, there is certainly much room for further investigation. We expect that the choice of RBN updating scheme will have little effect on the fundamentals of the phase transitions reported here (as per [18]), though this remains to be tested. Furthermore, we intend to explore the effect of different topologies on the phase transitions, in particular small-world (akin to [48])[9] and scale-free topologies (since most biological networks are scale-free with an exponent putting them near the critical point [50]). In a related fashion, we plan to examine the effect of altering the topology of power grids (e.g. with mini/microgrid technology [51]), and whether the insights gained here can be used to control such cascade events. The effect of topology is highlighted as a key step in understanding cascading effects in financial markets [52], and would be analogous to the investigation of altering edge weights in [53]. Also, the effect of noise on the information dynamics in the network should be established (in a similar fashion to its incorporation in the investigation of avalanche size distributions in [6]). Also requiring investigation is whether the information dynamics here can be used to drive evolution or self-tuning adaptation to produce critical networks, e.g. whether maximising information transfer between nodes can produce a preferential attachment effect as conjectured in Sect. 6.2.3. Such an experiment with RBNs could provide evidence that an underlying capacity for computation may have been a driver in GRN evolution. Finally, the framework should be applied to Boolean network or other models of particular real-world GRNs (e.g. [45]) in order to provide insight into the computations taking place there.[10]

References

1. M. Mitchell, *Complexity: A Guided Tour* (Oxford University Press, New York, 2009)
2. M. Mitchell, Complex systems: network thinking. Artif. Intell. **170**(18), 1194–1212 (2006)
3. R.V. Solé, S. Valverde, Information transfer and phase transitions in a model of internet traffic. Physica A **289**(3–4), 595–605 (2001)
4. R.V. Solé, S. Valverde, Information theory of complex networks: on evolution and architectural constraints, in *Complex Networks*, ed. by E. Ben-Naim, H. Frauenfelder, Z. Toroczkai. Lecture Notes in Physics, vol. 650 (Springer, Berlin, 2004), pp. 189–207
5. O. Kinouchi, M. Copelli, Optimal dynamical range of excitable networks at criticality. Nat. Phys. **2**(5), 348–351 (2006)
6. P. Rämö, S. Kauffman, J. Kesseli, O. Yli-Harja, Measures for information propagation in Boolean networks. Physica D **227**(1), 100–104 (2007)

[9] We have since made this exploration of small-world Boolean networks in [49].

[10] Note however that applying the framework to acyclic or feed-forward models such as [29, 45, 54, 55] (which are widely available since they are much simpler to infer) will reveal little more than trivial computation. This is because in the absence of external stimuli, acyclic models reach an attractor state very quickly, and the computation of the network is then completed.

7. A.S. Ribeiro, S.A. Kauffman, J. Lloyd-Price, B. Samuelsson, J.E.S. Socolar, Mutual information in random Boolean models of regulatory networks. Phys. Rev. E **77**(1), 011901 (2008)
8. V. Latora, M. Marchiori, Efficient behavior of small-world networks. Phys. Rev. Lett. **87**(19), 198701 (2001)
9. D. Polani, O. Sporns, M. Lungarella, How information and embodiment shape intelligent information processing, in *Proceedings of the 50th Anniversary Summit of Artificial Intelligence, New York*, ed. by M. Lungarella, F. Iida, J. Bongard, R. Pfeifer. ser. Lecture Notes in Computer Science, vol. 4850 (Springer, Berlin, 2007), pp. 99–111
10. C. Gershenson, Introduction to random Boolean networks, in *Proceedings of the Workshops and Tutorials of the Ninth International Conference on the Simulation and Synthesis of Living Systems (ALife IX), Boston, USA*, ed. by M. Bedau, P. Husbands, T. Hutton, S. Kumar, H. Suzuki, 2004, pp. 160–173
11. P. Fernández, R.V. Solé, The role of computation in complex regulatory networks, in *Scale-Free Networks and Genome Biology*, ed. by E.V. Koonin, Y.I. Wolf, G.P. Karev (Landes Bioscience, Georgetown, 2006), pp. 206–225
12. C.G. Langton, Computation at the edge of chaos: phase transitions and emergent computation. Phys. D **42**(1–3), 12–37 (1990)
13. S.A. Kauffman, *The Origins of Order: Self-Organization and Selection in Evolution* (Oxford University Press, New York, 1993)
14. O. Miramontes, Order-disorder transitions in the behavior of ant societies. Complexity **1**(3), 56–60 (1995)
15. D. Coffey, Self-organization, complexity and chaos: the new biology for medicine. Nat. Med. **4**(8), 882–5 (1998)
16. M. Mitchell, P.T. Hraber, J.P. Crutchfield, Revisiting the edge of chaos: evolving cellular automata to perform computations. Complex Syst. **7**, 89–130 (1993)
17. D.P. Feldman, C.S. McTague, J.P. Crutchfield, The organization of intrinsic computation: complexity-entropy diagrams and the diversity of natural information processing. Chaos **18**(4), 043106 (2008)
18. C. Gershenson, Updating schemes in random Boolean networks: do they really matter? in *Proceedings of the Ninth International Conference on the Simulation and Synthesis of Living Systems (ALife IX), Boston, USA*, ed. by J. Pollack, M. Bedau, P. Husbands, T. Ikegami, R.A. Watson (MIT Press, Cambridge, 2004), pp. 238–243
19. M. Mitchell, J. P. Crutchfield, and P. T. Hraber, Dynamics, computation, and the edge of chaos: a re-examination, in *Complexity: Metaphors, Models, and Reality*, ed. by G. Cowan, D. Pines, D. Melzner. Santa Fe Institute Studies in the Sciences of Complexity, vol. 19 (Addison Wesley, Reading, 1994), pp. 497–513
20. C. Gershenson, Phase transitions in random Boolean networks with different updating schemes. arXiv:nlin/0311008v1 (2004), http://arxiv.org/abs/nlin/0311008
21. J.T. Lizier, M. Prokopenko, A.Y. Zomaya, The information dynamics of phase transitions in random Boolean networks, in *Proceedings of the Eleventh International Conference on the Simulation and Synthesis of Living Systems (ALife XI), Winchester, UK*, ed. by S. Bullock, J. Noble, R. Watson, M.A. Bedau (MIT Press, Cambridge, 2008), pp. 374–381
22. J.T. Lizier, M. Prokopenko, D.J. Cornforth, The information dynamics of cascading failures in energy networks, in *Proceedings of the European Conference on Complex Systems (ECCS), Warwick, UK*, 2009, p. 54, ISBN: 978-0-9554123-1-8
23. C. Gershenson, RBNLab, Online software (2003), http://rbn.sourceforge.net
24. B. Derrida, Y. Pomeau, Random networks of automata: a simple annealed approximation. Europhys. Lett. **1**(2), 45–49 (1986)
25. D.P. Feldman, J.P. Crutchfield, Discovering noncritical organization: Statistical mechanical, information theoretic, and computational views of patterns in one-dimensional spin systems, Santa Fe Institute Working Paper 98-04-026 (1998), http://www.santafe.edu/media/workingpapers/98-04-026.pdf
26. P. Crucitti, V. Latora, M. Marchiori, Model for cascading failures in complex networks. Phys. Rev. E **69**(4), 045104 (2004)

27. A.E. Motter, Y.-C. Lai, Cascade-based attacks on complex networks. Phys. Rev. E **66**(6), 065102 (2002)
28. P. Ormerod, A. Heineike, Global recessions as a cascade phenomenon with interacting agents. J. Econ. Interact. Coord. **4**(1), 15–26 (2009)
29. P. Fernández, R.V. Solé, Neutral fitness landscapes in signalling networks. J. Roy. Soc. Interface **4**(12), 41–47 (2007)
30. I. Couzin, R. James, D. Croft, J. Krause, Social organization and information transfer in schooling fishes, in *Fish Cognition and Behavior*, ed. by B.C.K. Laland, J. Krause. Fish and Aquatic Resources (Blackwell Publishing, Oxford, 2006), pp. 166–185
31. D. Peak, J.D. West, S.M. Messinger, K.A. Mott, Evidence for complex, collective dynamics and emergent, distributed computation in plants. Proc. Natl. Acad. Sci. USA. **101**(4), 918–922 (2004)
32. K.I. Goh, A.L. Barabási, Burstiness and memory in complex systems. Europhys. Lett. **81**(4), 48002 (2008)
33. R. Kinney, P. Crucitti, R. Albert, V. Latora, Modeling cascading failures in the north american power grid. Eur. Phys. J. B **46**(1), 101–107 (2005)
34. Y. Xia, J. Fan, D. Hill, Cascading failure in Watts-Strogatz small-world networks. Physica A **389**(6), 1281–1285 (2010)
35. K.I. Goh, B. Kahng, D. Kim, Universal behavior of load distribution in scale-free networks. Phys. Rev. Lett. **87**(27), 278701 (2001)
36. D.J. Watts, S. Strogatz, Collective dynamics of 'small-world'networks. Nature **393**, 440–442 (1998)
37. A. Tang, C. Honey, J. Hobbs, A. Sher, A. Litke, O. Sporns, J. Beggs, Information flow in local cortical networks is not democratic. BMC Neurosci. **9**(1), O3 (2008)
38. C.J. Honey, R. Kotter, M. Breakspear, O. Sporns, Network structure of cerebral cortex shapes functional connectivity on multiple time scales. Proc. Natl. Acad. Sci. **104**(24), 10 240–10 245 (2007)
39. M. Piraveenan, M. Prokopenko, A.Y. Zomaya, Local assortativeness in scale-free networks. Europhys. Lett. **84**(2), 28002 (2008)
40. M. Piraveenan, M. Prokopenko, A.Y. Zomaya, Local assortativity and growth of internet, Eur. Phys. J. B **70**(2), 275–285 (2009)
41. E. Estrada, Information mobility in complex networks. Phys. Rev. E **80**(2), 026104 (2009)
42. A.-L. Barabási, R. Albert, Emergence of scaling in random networks. Science **286**(5439), 509–512 (1999)
43. A.-L. Barabási, R. Albert, H. Jeong, Scale-free characteristics of random networks: The topology of the world-wide web. Physica A **281**, 69–77 (2000)
44. M. Prokopenko, F. Boschietti, A.J. Ryan, An information-theoretic primer on complexity, self-organization, and emergence. Complexity **15**(1), 11–28 (2009)
45. A. Samal, S. Jain, The regulatory network of e. coli metabolism as a Boolean dynamical system exhibits both homeostasis and flexibility of response. BMC Syst. Biol. **2**(1), 21 (2008)
46. P. Rämö, J. Kesseli, O. Yli-Harja, Perturbation avalanches and criticality in gene regulatory networks. J. Theor. Biol. **242**(1), 164–170 (2006)
47. I. Couzin, Collective minds. Nature **445**(7129), 715–715 (2007)
48. Q. Lu, C. Teuscher, Damage spreading in spatial and small-world random Boolean networks. arXiv:0904.4052 (2009), http://arxiv.org/abs/0904.4052
49. J.T. Lizier, S. Pritam, M. Prokopenko, Information dynamics in small-world Boolean networks. Artif. Life **17**(4), 293–314 (2011)
50. M. Aldana, Boolean dynamics of networks with scale-free topology. Physica D **185**(1), 45–66 (2003)
51. S. Abu-Sharkh, R.J. Arnold, J. Kohler, R. Li, T. Markvart, J.N. Ross, K. Steemers, P. Wilson, R. Yao, Can microgrids make a major contribution to UK energy supply? Renew. Sustain. Energy Rev. **10**(2), 78–127 (2006)
52. F. Schweitzer, G. Fagiolo, D. Sornette, F. Vega-Redondo, A. Vespignani, D.R. White, Economic networks: The new challenges. Science **325**(5939), 422–425 (2009)

53. R. Yang, W.-X. Wang, Y.-C. Lai, G. Chen, Optimal weighting scheme for suppressing cascades and traffic congestion in complex networks. Phys. Rev. E **79**(2), 026112 (2009)
54. M.W. Covert, E.M. Knight, J.L. Reed, M.J. Herrgard, B.O. Palsson, Integrating high-throughput and computational data elucidates bacterial networks. Nature **429**(6987), 92–96 (2004)
55. M.I. Davidich, S. Bornholdt, Boolean network model predicts cell cycle sequence of fission yeast. PLoS ONE **3**(2), e1672 (2008)

Chapter 7
Coherent Information Structure in Complex Computation

The framework for the information dynamics of distributed computation introduced in Chaps. 3–5 has proven successful in locally identifying the component operations of information storage, transfer and modification. We have observed that while these component operations exist to some extent in all types of computation, complex computation is distinguished in having *coherent structure* in its local information dynamics profiles. We *conjecture* that *coherent information structure* is a defining feature of complex computation, particularly evolved computation that solves human-understandable tasks. We present a methodology for studying coherent information structure, consisting of state-space diagrams of the local information dynamics and a measure of structure in these diagrams.[1] The methodology identifies both clear and "hidden" coherent structure in complex computation, most notably reconciling conflicting interpretations of the complexity of ECA rule 22. The measure is also used to demonstrate a maximisation of coherent information structure in the order-chaos phase-transition in RBNs.

7.1 Introduction

Coherent information structure is a feature that is consistently observed in complex computation. In particular it appears that *nature evolves* coherent computation. Illustrative examples include: coherent signalling cascades providing information transfer in gene networks [2, 3]; coherent waves of directional change in flocking birds [4] which are also referred to as "information cascades" in schooling fish in [5]; the dynamic process of opening and closing stomatal apertures in plants via coherent collective waves of activity [6]; and coherent wave structures in neural computation [7]. Indeed, we observe that coherent computational structure also emerges from evolution in *artificial* systems, e.g. the ϕ_{par} CA rule which was evolved to solve

[1] The methodology for studying coherent information structure and the results of its application to CAs were first reported in [1].

J. T. Lizier, *The Local Information Dynamics of Distributed Computation in Complex Systems*, Springer Theses, DOI: 10.1007/978-3-642-32952-4_7,

© Springer-Verlag Berlin Heidelberg 2013

the density classification problem by using coherent glider structures [8, 9] (see Fig. 3.6).[2]

In our own work, we have qualitatively observed in Chaps. 3–5 that the known complex ECA rules (54 and 110) exhibit the largest amount of coherent information structure. That is, they contain much emergent coherent structure (gliders, blinkers and domain walls) in the local profiles in space and time of their information dynamics: information storage, transfer and modification. This observation aligns well with similar explorations of other types of local information structure for these rules, e.g. [10–12]. Indeed, one must note that all of the above examples of coherent structure in biological systems are analogous to gliders in CAs. Also, we have demonstrated maximisations of the information storage and single-source transfer near the critical state in order-chaos phase transitions in RBNs in Sect. 6.1. There we suggested the results could be interpreted as a maximum capacity for coherent computation near the critical state.

On the basis of the above observations, we *conjecture* that the *coherence* of local information structure is a defining feature of complex computation, particularly evolved computation which solves human-understandable tasks. "Coherence" implies a property of sticking together or a logical relationship [13]: in this context we use the term to qualitatively describe **a logical spatiotemporal relationship between values in local information dynamics profiles**. For example, the manner in which particles in CAs give rise to similar values of local information transfer amongst spatiotemporal neighbours is coherent in this sense.[3] We emphasise it is the *information structure* that is coherent here: the original CA states (from which the presence of particles is not clear) are not always obviously coherent themselves.

Using language reminiscent of Langton's analysis [14], we suggest that complex systems exhibit very *highly-structured coherent* computation in comparison to:

a. ordered systems, which exhibit coherence but minimal structure in a computation dominated by information storage; and
b. chaotic systems, whose computations are dominated by rampant information transfer eroding any coherence.

Coherent structure is *useful* in complex computation because it provides *stable* mechanisms for storing information, transferring information, and modifying that information when required. Presumably it emerges in evolution of complex computation because of this utility. For these reasons, we suggest that coherent information structure may be a useful intrinsic goal in the domain of *guided self-organisation* [15]. Evolution of coherent information structure could be particularly useful where task-based evolution faces initially flat task-based fitness landscapes, perhaps serving as a platform from which to launch better-equipped task-based evolution. Furthermore, coherent information structure is associated with computation

[2] Also, in Sect. 8.3 we will observe the emergence of coherent particle-like information structures in a snake-like robot evolved to maximise transfer entropy between its segments.

[3] See also [10] for a discussion of the term "coherent structure" referring to particles (including blinkers) in this context.

that appears relatively simple for humans to understand (e.g. the ϕ_{par} CA), which is an important trait for both acceptance and maintenance of self-organised systems in real-world deployments.

Our *goal* then is to measure the coherence of computational or information structure in a given system from its local information dynamics. That is, given a system $\mathbf{X}$ (i.e. with a time series $x_{i,n}$ for each variable X_i in the system, and a set of causal connections between them), we wish to compute a measure of the coherence of information structure in it. The approach is intended to first compute the local information dynamics at each space-time point in the system, resulting in the individual profiles we have seen in Chaps. 3–5. The next step is to compute the measure of coherent information structure from these profiles. Existing measures for coherent structure are generally problem-specific and consider only contiguous clusters of similar values. A typical example is [16] which uses a mix of thresholding and spatiotemporal clustering of resulting binary values to identify specific coherent structures. Here however, we seek a general measure portable between different types of computation. Certainly the examination of local information dynamics values provides a system-independent perspective. In addition to this, measuring coherence here must take into account the *continuous* nature of the information dynamics measure, the fact that there are *multiple* information dynamics measures, and be *generic* in detecting coherent structure as relationships *between* these measures that may be more subtle than a contiguous cluster of similar values in a profile of one of them.

We also note that the goal of measuring coherent information structure is distinct from attempting to measure complexity itself (e.g. with the statistical complexity [10, 17]), since it is a particular concept or system property. Indeed, it is unclear whether overall complexity and coherent computational structure will have a one-to-one correspondence. Guiding self-organisation towards generally complex computation may not produce the coherent structure that we argued above to be useful. Also, there are examples of computation argued to be complex from certain perspectives despite the apparent absence of local coherent structure. A prominent example is CA rule 22 (see arguments in favour of its complexity in Sect. 2.3.4), where these conflicting perspectives are yet to be reconciled. Put simply, the concept of coherent information structure is worth exploring in its own right. On a related note though, we will also examine whether measuring coherent information structure provides useful insights into the differences in complexity of various computations. As outlined above, we would expect the measure to be large in known complex computation. We emphasise though that while this measure is not expected to produce a *universal* complexity-entropy curve (as per [18]), it could be expected to be maximised along with known complex behaviour in certain simple order-chaos phase transitions, e.g. in RBNs.

In this chapter, we focus on measuring coherent information structure in CAs and RBNs. In Appendix G we demonstrate that several obvious candidate measures do not meet the requirements we laid out above. A prominent example is the apparent transfer entropy itself, which is a candidate since it measures coherent effects of a single source on a destination, large local values of it are associated with coherent glider structures in CAs, and it was shown to be maximised near the critical state in the phase transition in RBNs. As shown in Appendix G however, there is not a one-to-one correspondence in any given CA between large average values of this measure and the existence of gliders or complex behaviour in general. As such, this measure itself cannot be considered to directly capture coherent information structure. Despite this, these candidate measures do provide further circumstantial evidence for the importance of coherent information structure in complex computation.

In Sect. 7.2 then, we present and explore diagrams of the multi-dimensional state-space formed by the local information dynamics, in particular the local information storage values and local values of information transfer from each source to a destination. This perspective allows the most general, unbiased interpretation of coherent information structure as a logical relationship between the information dynamics, in alignment with our qualitative interpretation and requirements above. Importantly, we demonstrate these diagrams as a particularly useful tool in this context. We conclude that coherent computational structure should be measured as *a logical relationship between values in the local information dynamics state-space*, and present the measure I^{ss} for this purpose in Sect. 7.3. The state-space diagrams and the measure I^{ss} provide a methodology which identifies clear and "hidden" coherent structure in complex computation, in particular revealing previously hidden coherent structure in the information dynamics of rule 22 and reconciling the differing perspectives on how complex it is. Finally, we demonstrate in Sect. 7.3.2 that the measure of coherent structure is maximised near the order-chaos phase transition in RBNs, as expected.

7.2 Local Information Dynamics State-Space

As per our qualitative description, coherence may be broadly interpreted as a logical relationship *between* the individual local information dynamics measures rather than only within individual profiles alone. Indeed, using a more broad definition allows an unbiased measurement of coherent structure. To explore relationships as per this broad definition, in this section we investigate information state-space diagrams which plot the local information dynamics against each other. We demonstrate that these diagrams are useful tools in revealing both clear and hidden coherent information structure in the computation in CAs.

The state-space referred to here is the multi-dimensional information space consisting of the local values of each of the individual information dynamics. In theory, plots of this state-space should reveal any logical relationship between the individual measures, and by our broad definition such relationships embody coherent information structure. Figure 7.1 projects the multi-dimensional information state-space onto two-dimensional diagrams. There we plot the local apparent transfer entropy $t(i, j = 1, n, k = 16)$ versus local active information storage $a(i, n, k = 16)$, as well as the local separable information $s(i, n, k = 16)$ versus $a(i, n, k = 16)$ for several

CA rules. Each point in these diagrams represents the local values of the measures at one spatiotemporal point (i, n). We emphasise that coherent *spatio*temporal structure can be captured in these diagrams since the transfer entropy measurements consider neighbouring values across space and the temporal history of the destination.

Similar state-space diagrams are known to provide insights into structure that are not visible when examining either measure in isolation. An example is examining structure in classes of systems (such as logistic maps) by plotting average excess entropy versus entropy rate while changing a system parameter [18]. In contrast here we are looking at structure *within* a single system rather than across a class of systems.

As could be expected, **the state-space diagrams for rule 110 exhibit interesting structure** (Fig. 7.1a and b), with significant clustering around certain areas and lines, reflecting its status as a complex rule. The two diagonal lines in Fig. 7.1a are upper limits representing the boundary condition $t^c(i, j = -1, n, k = 16) \geq 0$ for both destination states "0" and "1". That is, via Eq. (4.57) we have:

$$h(i, n+1) \geq a(i, n+1, k) + t(i, j = 1, n+1, k), \quad (7.1)$$

for both the upper limits $h(i, n+1) = -\log_2 p(x_{i,n+1} = 0)$ and $-\log_2 p(x_{i,n+1} = 1)$. The horizontal line is for $t(i, j = 1, n, k = 16) = 0$. Rather than being merely boundaries these lines are structural elements in their own right, being dense areas of the state-space and adding significant structure to it. Their long extent suggests a large variety of coherent activity, because with zero transfer from one source and $a(i, n+1, k)$ much less than 0 the other source has a greater opportunity to exert a strong coherent influence on the destination.

We then consider the chaotic rule 30, whose local information dynamics profiles in Fig. 3.11 showed a similar absence of coherent structure to rule 22 (see Sect. 3.3.8). In contrast to the results for rule 110, **the state-space diagrams for rule 30 exhibit minimal structure** (Fig. 7.1c and d), with a smooth spread of points across the space reflecting its underlying chaotic nature. The same mathematical limits exist of course but are not a dense, individual part of the structure as for rule 110.

As shown in Fig. 3.10, rule 22 exhibits an apparent absence of coherent structure in its individual space-time information profiles, similar to those described for rule 30. As such, one may expect state-space diagrams for rule 22 to exhibit a similar absence of structure. As shown by Fig. 7.1e and f however this is not the case: **the state-space diagrams for rule 22 exhibit significant structure**, with similar clustering to that of rule 110. Indeed, rule 22 exhibits similar linear structures to rule 110, which as argued above suggests the potential for strong coherent influence of a source on a destination.

This is a *highly significant result*, because **local information dynamics is the first approach able to reconcile the opposing views of rule 22 as complex and chaotic.**[4] When the local information dynamics profiles are viewed

[4] Indeed, the candidate measures considered in Appendix G did not capture its alignment with the known complex rules in this respect.

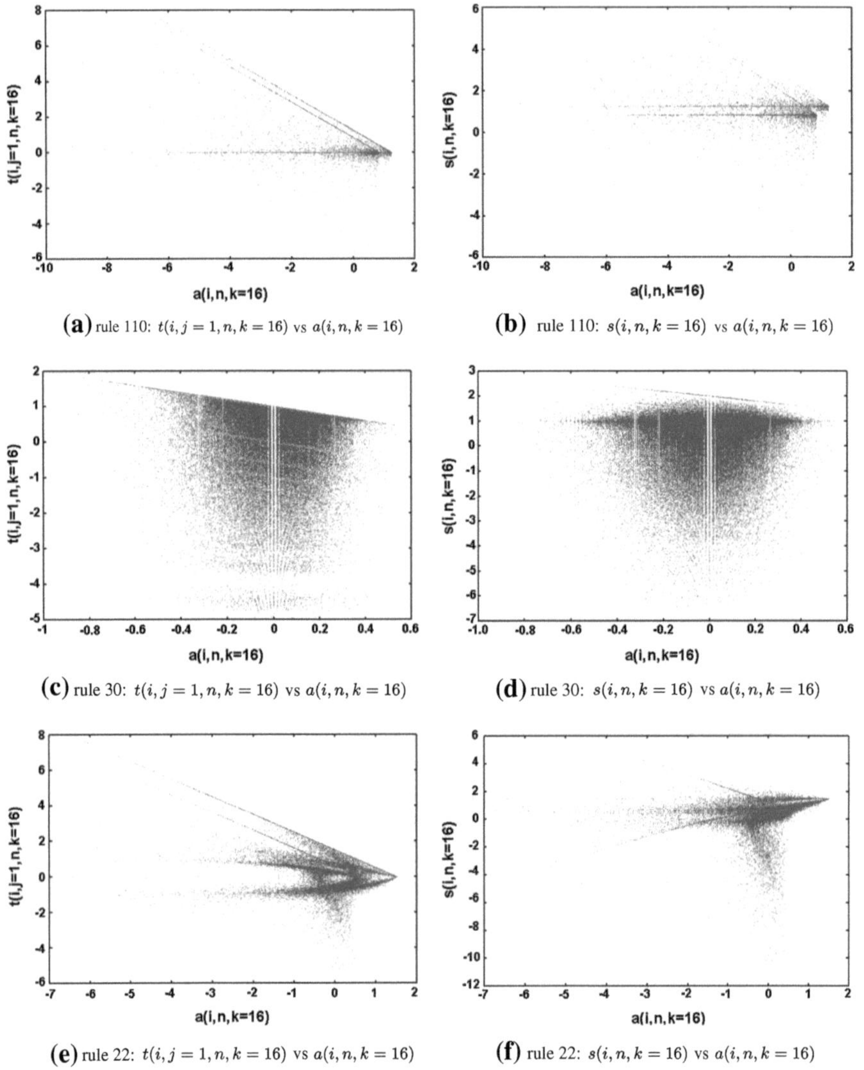

(a) rule 110: $t(i, j = 1, n, k = 16)$ vs $a(i, n, k = 16)$ **(b)** rule 110: $s(i, n, k = 16)$ vs $a(i, n, k = 16)$

(c) rule 30: $t(i, j = 1, n, k = 16)$ vs $a(i, n, k = 16)$ **(d)** rule 30: $s(i, n, k = 16)$ vs $a(i, n, k = 16)$

(e) rule 22: $t(i, j = 1, n, k = 16)$ vs $a(i, n, k = 16)$ **(f)** rule 22: $s(i, n, k = 16)$ vs $a(i, n, k = 16)$

Fig. 7.1 State space diagrams of local transfer entropy (one step to the *right*) $t(i, j = 1, n, k = 16)$, local active information $a(i, n, k = 16)$ and local separable information $s(i, n, k = 16)$ at the same space-time points (i, n) for ECA rules 110, 30 and 22. (NB: Fig. 7.1 **a**, **c** and **e** are reprinted from [1] with permission of Springer.)

individually (Fig. 3.10), no coherent structure is revealed. This is in alignment with local information profiles from other authors, e.g. [10]. In contrast, here we see that when plotted *together* in state-space diagrams we see an indication of a coherent relationship *between* the local information dynamics. This view lends credence to the claims of complex behaviour for rule 22 discussed in Sect. 2.3.4. Its coherent information structure must be very subtle, since it is not complex enough to be revealed in the individual profiles. Indeed, this structure may underpin coherent computation at other scales.

7.3 Measuring Coherent Information Structure in the State-Space

Coherent information structure should be quantified from these state-space diagrams, because they satisfy the requirements described in Sect. 7.1 and have been demonstrated in Sect. 7.2 to identify both clear and hidden structure. As such, we conclude that **coherent information structure should be quantified as the logical relationship between values in the local information dynamics state-space**. In this section, we discuss how to quantify it in this manner and present the measure I^{ss} for this purpose. Subsequently, we apply the measure I^{ss} to CAs and RBNs and discuss the results.

To remain maximally unbiased, the measurement should be made in the underlying multi-dimensional information state-space rather than the two-dimensional projections plotted in Fig. 7.1. The first question to address is which of the information dynamics should be included in the measure here. Certainly, all of the information sources for a given destination must be represented. One could do so by examining the state-space made up of $a_X(n,k)$ and the incrementally conditioned mutual information terms in Eq. 4.47 (e.g. $a(i,n,k)$, $t(i,j=1,n,k)$, $t^c(i,j=-1,n,k)$ in ECAs). However, this approach leaves ambiguity in the order of considering sources in the incrementally conditioned mutual information terms. Also, as more sources are conditioned on, more redundant information is built into this state-space about the previously considered sources.[5] A more appropriate approach without such ambiguity or inbuilt bias is to consider the state-space of measures underlying the separable information: i.e. $a_X(n,k)$ and $t_{Y\to X}(n,k)$ for all sources Y. The separable information should not be included in the measure itself since it includes redundant information about its component variables.

The next question is how to measure structure in the state-space of these measures. Certainly, measuring structure in two or more dimensional patterns is known to be particularly difficult [19]. We select the multi-information (see Eq. 2.9) for this purpose, since it measures the degree of dependence between the given variables. This aligns well with our intention to measure the *logical relationship* between the information dynamics variables.

As such, we propose the **state-space multi-information** as a measure of *coherent information structure* at an information destination X due to causal sources $\{\mathbf{V}_X \setminus X\}$:

[5] For example, $t^c(i,j=-1,n,k)$ is almost completely specified by $a(i,n,k)$ and $t(i,j=1,n,k)$ in ECAs, except for any difference in $h(i,n)$ between the "0" and "1" states.

$Y_1, Y_2, ..., Y_G$ as the multi-information I_X^{ss} between its active information storage and the apparent transfer entropies to it[6]:

$$I_X^{ss} = \lim_{k\to\infty} I_X^{ss}(k), \tag{7.2}$$

$$I_X^{ss}(k) = I\left(a_X(n,k); t_{Y_1\to X}(n,k); t_{Y_2\to X}(n,k); \ldots; t_{Y_G\to X}(n,k)\right), \tag{7.3}$$

$$= H\left(a_X(n,k)\right) + \left(\sum_{g=1}^{G} H\left(t_{Y_g\to X}(n,k)\right)\right)$$
$$- H\left(a_X(n,k); t_{Y_1\to X}(n,k); t_{Y_2\to X}(n,k); \ldots; t_{Y_G\to X}(n,k)\right). \tag{7.4}$$

For a lattice system we can write $I^{ss}(i,k)$ for agent i; where the agents are homogeneous (e.g. CAs) we can estimate the PDFs over all observations in the system, and average over all agents to get $I^{ss}(k) = \langle I^{ss}(i,k)\rangle_i$. Initially, it may seem strange to make an information-theoretic measure of information-theoretical values; however there is no reason that structure in the information state-space should not be measured information-theoretically.

7.3.1 Coherent Information Structure Measurements in CAs

The state-space multi-information $I^{ss}(k=16)$ was measured for the CA rules investigated in previous chapters. These measurements used 200,000 sample points, with kernel estimation of the underlying entropies (see Sect. 2.2.3, [20]) since the sample points have continuous values. These estimates use a step kernel with precision $r = 1.0$ bits and maximum distance norms. The results are displayed in Table 7.1.

The measurements of coherent information structure using $I^{ss}(k=16)$ align to a large degree with our qualitative observations of these rules above. We find that rules 110 and 54 contain strong relationships between the information dynamics in their state-space. In contrast, rule 30 contains very little coherent structure. We also measure only a small amount of coherent structure for rule 18, since the structure in its domain walls is only a relatively minor part of the dynamics compared to its background domain. While there is a pattern to this domain, as we have observed there is almost no relationship between the interacting sources (see Sect. 4.2.4.2), so the lack of coherent structure here is not surprising. As per our qualitative observations above, significant coherent structure is also measured for rule 22. This confirms that the framework can resolve the conflicting views of rule 22 as complex and chaotic. We remind the reader that this measure is not of complexity itself and, given the

[6] Note we have altered our notation for mutual and multi-information expressions from $I_{X;Y}$ to $I(X;Y)$ here. This is simply for easier display of the complicated quantities over which the information is being calculated.

subtle nature of structure of the coherent structure in rule 22 (not being revealed in *individual* information dynamics profiles), at this stage we cannot conclude that its behaviour is any more complex than that of rules 110 and 54.

We also note that the measure returns large values for CA rules (e.g. 0.90 for rule 47) which contain simple gliders in a single channel j only. Since these gliders never interact with other gliders, these rules are not viewed as complex. The large measurement is not incorrect though, since the measure is genuinely detecting coherent structure. We note that this structure only manifests in a single two-dimensional projection of the state-space ($t(i, j, n, k)$ versus $a(i, n, k)$ for the channel j the gliders travel in). It is undesirable to infer as much coherent structure to these types of dynamics as for rules such as 110 which have structure in all dimensions of the state-space. We expect that new approaches addressing the acknowledged difficulties in defining measures of multi-dimensional structure (e.g. see [19]) may provide improvements over the multi-information for measuring structure in the local information state-space here.

Despite this minor limitation, these results (in particular for rule 22) underline that the measure I^{ss} usefully captures clear and hidden coherent information structure observed in the state-space. We emphasise though, that the state-space diagrams themselves contain much more detail about the relationships between these axes of complexity than this averaged measure.[7]

7.3.2 *Coherent Information Structure Measurements in RBNs*

We also applied the measure I^{ss} to RBNs, seeking to measure the average $I^{ss}(\overline{K})$ as a function of average in-degree or connectivity $\overline{K}$. Given that a phase transition occurs between ordered behaviour at low $\overline{K}$ and chaotic behaviour at high $\overline{K}$, one could expect to measure a maximisation of coherent information structure coinciding with complex behaviour at the critical regime.

The RBNs are simulated using the same parameters as for the experiments in Sect. 6.1, except where noted below. Again, because we use finite network sizes, the phase transitions investigated here are approximate (see Sect. 2.4.1). We measure the local information dynamics for each node in a given sample RBN (e.g. $a_X(i, n, k = 13)$), then use these to compute the coherent information structure $I_X^{ss}(k = 13)$ for each node X. Note that the nodes are heterogeneous, in particular in terms of the number of causal information sources, so the information dynamics and state-space multi-information must be computed for each node separately. In order to generate enough observations for the underlying information dynamics computations, each sample RBN is run from a large number of random initial states (at least 8,960 are used). The measurements of $I_X^{ss}(k = 13)$ used the local information dynamics values

[7] This is similar to the manner in which the local information dynamics measures themselves reveal more about the underlying computation than their averages do.

Table 7.1 Table of the multi-information $I^{ss}(k = 16)$ in the local information state-space in ECAs

ECA rule	$I^{ss}(k = 16)$ (bits)
110	0.50
54	0.95
22	0.72
18	0.15
30	0.11

The state-space consists of the local active information storage ($a(i, n, k = 16)$) and each local transfer entropy ($t(i, j = 1, n, k = 16)$ and $t(i, j = -1, n, k = 16)$)

from at least 35,000 sample points. Kernel estimation is used for the underlying entropies since the sample points have continuous values (see Sect. 2.2.3, [20]) using a step kernel with precision $r = 0.4$ bits and maximum distance norms. The measures of $I_X^{ss}(k = 13)$ for each node X in the RBN then provide an average across the sample network $I^{ss}(k = 13) = \langle I_X^{ss}(k = 13) \rangle_X$. These sample network averages are then averaged over many networks generated for each $\overline{K}$ (at least 120) to determine the state-space multi-information as a function of $\overline{K}$: $I^{ss}(k = 13, \overline{K})$.

The results for $I^{ss}(k = 13, \overline{K})$ as a function of $\overline{K}$ are displayed in Fig. 7.2. $I^{ss}(k = 13, \overline{K})$ is maximised itself in between the maximisations of information storage and transfer, which we know to lie on either side of the phase transition (see Sect. 6.1). Clearly then, **our measure I^{ss} infers that the coherence of information structure is maximised directly in the vicinity of the critical phase between ordered and chaotic behaviour of the RBNs**. This is as we expected, given our earlier comments about phase transitions in general and the circumstantial evidence of such a maximisation inferred in Sect. G.1. In the ordered phase, the system exhibits coherence but with only minimal structure, while in the chaotic phase the coherence of the system is eroded by rampant interaction-based information transfer.

One could also extend the mutual information interpretation of [21, 22] to the multi-information expression here Eq. (7.4). In the ordered phase there is low assortative noise between the individual information dynamics, though since there is low diversity in each of them, coherence remains low. In the chaotic phase, there is high diversity in the individual information dynamics, but at the same time high assortative noise between them erodes the coherence. In the critical phase in between, there is large information storage and information transfer with high diversity in each providing a large scope for coherent information structure. Unlike the chaotic phase, here there is low assortative noise between them, and coherent information structure is then high in the critical phase.

We note that the minor limitation of I^{ss} described for CAs in Sect. 7.3.1 is not an issue in RBNs. Since RBNs are neither a lattice system nor consist of homogeneous agents, the existence of simple gliders in one channel is not meaningful.

While this finding of a maximisation of coherent information structure in this order-chaos phase transition is a significant result, there are a number of key points to keep in mind. The result is only for this specific phase transition. Certainly systems

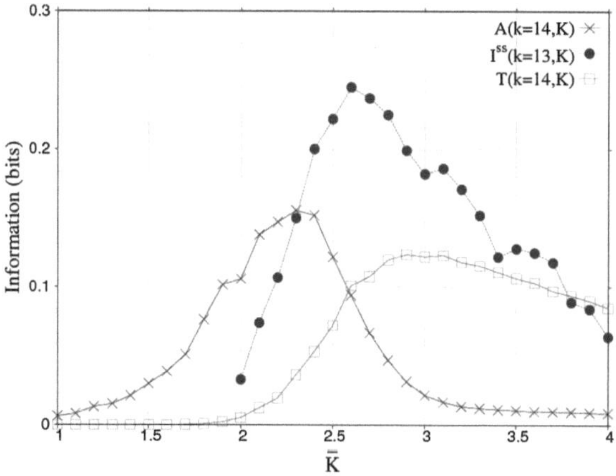

Fig. 7.2 Average coherent information structure $I^{ss}(k = 13, \overline{K})$ in the state-space diagrams for RBNs, as a function of average in-degree $\overline{K}$. $I^{ss}(k = 13, \overline{K})$ is plotted against the results of $A(k = 13, \overline{K})$ and $T(k = 13, \overline{K})$ from Sect. 6.1 for comparison. Standard error of the mean indicators are on the scale of the data points and omitted

near the critical phase displayed the largest amount of coherent structure on average, but such structure can exist to an extent in all types of systems. While there are good reasons to expect similar trends in other known phase transitions where complexity is maximised, certainly there is no one universal complex-entropy curve [18]. Indeed, we remind the reader that this measure of coherent information structure is expected to capture an important feature of complex computation, but is not an overall measure of complexity itself. Also, we emphasise that there is a large range of possible behaviour for each $\overline{K}$ in the finite-size networks (hence the transition is approximate), and what we have reported here are average behaviours. The local information dynamics of computation will provide more detailed insights *for a given* RBN than single ensemble measures such as $I^{ss}(k = 13, \overline{K})$. Importantly though, these ensemble measures are useful guides, and indeed become much more strongly predictive of behaviour in the limit of infinite system size (e.g. see sharpening of phase transition with system size, and discontinuity at infinite size, for RBNs in [23, 24]).

7.4 Summary

We have conjectured that coherent information structure is a defining feature of complex distributed computation, particularly in biological systems. We have presented a methodology for exploring the coherence of such computation, using our framework for local information dynamics. The methodology focuses on state-space diagrams of the local information dynamics, since these provide the most unbiased approached to detecting logical relationships between the local information dynamics values.

These diagrams produce useful visual insights and indeed we define coherent information structure as the logical relationship between values in the local information dynamics state-space. We proposed the multi-information I^{ss} in this state-space as a measure for this concept.

We applied the methodology to classical complex systems examples: cellular automata and random Boolean networks. These applications showed that the method captures not only clear but also hidden coherent structure in the underlying computation. A prominent example here is CA rule 22, where no coherent structure is visible in spatiotemporal profiles of the individual information dynamics, but is revealed in the state-space diagrams between them and quantified by our measure I^{ss}. This is a significant result, since our framework is the first approach which can reconcile the two conflicting views of rule 22 as complex and chaotic. On application to RBNs, our measure I^{ss} suggests that coherent information structure is maximised at the critical state in the order-chaos phase transition there. While we emphasise that I^{ss} is not intended to measure complexity as a whole, this result provides important evidence that coherent information structure is a key feature of complex computation.

We intend to explore the relationship of coherent information structure to other properties of complex computation, and indeed other measures of complexity. In particular, further work is required to establish the meaning of the relationship between the information dynamics measures in rule 22. We also intend to explore the application of new measures of multi-dimensional structure to the information state-space as they are published, in order to check if they can address the minor limitations of I^{ss} described in Sect. 7.3.1. Furthermore, we intend to explore the utility of the measure of coherent information structure I^{ss} as an intrinsic goal in *guided self-organisation*. In this domain, guiding towards higher values of I^{ss} has the potential to generate coherent computational structure which could be usefully applied to solve complex tasks.

References

1. J.T. Lizier, M. Prokopenko, A.Y. Zomaya, Coherent information structure in complex computation, Theory Biosci. **131**(3), 193–203 (2012), doi:10.1007/s12064-011-0145-9
2. P. Fernández, R.V. Solé, The role of computation in complex regulatory networks, in *Scale-free Networks and Genome Biology*, ed. by E.V. Koonin, Y.I. Wolf, G.P. Karev (Landes Bioscience, Georgetown, 2006), pp. 206–225
3. P. Fernández, R.V. Solé, Neutral fitness landscapes in signalling networks. J. R. Soc. Interface, **4**(12) 41–47 (2007)
4. A. Cavagna, A. Cimarelli, I. Giardina, G. Parisi, R. Santagati, F. Stefanini, M. Viale, Scale-free correlations in starling flocks, in *Proceedings of the National Academy of Sciences*, vol. 107, no. 26, pp. 11, 865–11,870 (2010)
5. I. Couzin, R. James, D. Croft, J. Krause, Social Organization and Information Transfer in Schooling Fishes, in *Fish Cognition and Behavior, ser. Fish and Aquatic Resources*, ed. by B.C.K. Laland, J. Krause (Blackwell Publishing, Oxford, 2006), pp. 166–185

6. D. Peak, J.D. West, S.M. Messinger, K.A. Mott, Evidence for complex, collective dynamics and emergent, distributed computation in plants. Proc. Nat. Acad. Sci. U.S.A. **101**(4), 918–922 (2004)
7. P. Gong, C. van Leeuwen, Distributed dynamical computation in neural circuits with propagating coherent activity patterns. PLoS Comput. Biol. **5**(12), e1000611 (2009)
8. M. Mitchell, J.P. Crutchfield, P.T. Hraber, Evolving cellular automata to perform computations: Mechanisms and impediments. Phys. D. **75**, 361–391 (1994)
9. M. Mitchell, J. P. Crutchfield, and R. Das, Evolving cellular automata with genetic algorithms: A review of recent work, in *Proceedings of the First International Conference on Evolutionary Computation and Its Applications* ed. by E.D. Goodman, W. Punch, V. Uskov. (Russian Academy of Sciences, Moscow, 1996)
10. C.R. Shalizi, R. Haslinger, J.-B. Rouquier, K.L. Klinkner, C. Moore, Automatic filters for the detection of coherent structure in spatiotemporal systems. Phys. Rev. E **73**(3), 036104 (2006)
11. T. Helvik, K. Lindgren, and M. G. Nordahl, Local information in one-dimensional cellular automata, in *Proceedings of the International Conference on Cellular Automata for Research and Industry*, ed. by P.M. Sloot, B. Chopard, A.G. Hoekstra. Amsterdam, ser. Lecture Notes in Computer Science. vol. 3305. (Springer, Berlin, 2004), pp. 121–130
12. J.E. Hanson, J.P. Crutchfield, The attractor-basin portait of a cellular automaton. J. Stat. Phys. **66**, 1415–1462 (1992)
13. Oxford english dictionary(2008) [Online]. Available url: http://www.oed.com/ Accessed 8 May 2008
14. C.G. Langton, Computation at the edge of chaos: phase transitions and emergent computation. Physica D **42**(1–3), 12–37 (1990)
15. M. Prokopenko, Guided self-organization. HFSP J. **3**(5), 287–289 (2009)
16. P. Jung, J. Wang, R. Wackerbauer, K. Showalter, Coherent structure analysis of spatiotemporal chaos. Phys. Rev. E. **61**(2), 2095–2098 (2000)
17. J.P. Crutchfield, K. Young, Inferring statistical complexity. Phys. Rev. Lett. **63**(2), 105 (1989)
18. D.P. Feldman, C.S. McTague, J.P. Crutchfield, The organization of intrinsic computation: Complexity-entropy diagrams and the diversity of natural information processing. Chaos **18**(4), 043106 (2008)
19. D.P. Feldman, J.P. Crutchfield, Structural information in two-dimensional patterns: Entropy convergence and excess entropy. Phys. Rev. E. **67**(5), 051104 (2003)
20. H. Kantz, T. Schreiber, *Nonlinear Time Series Analysis* (Cambridge University Press, Cambridge, 1997)
21. R. V. Solé and S. Valverde, Information theory of complex networks. On evolution and architectural constraints, in *Complex Networks*, ser. eds. E. Ben-Naim, H. Frauenfelder, Z. Toroczkai. Lecture Notes in Physics, vol. 650, (Springer, Berlin, 2004) pp. 189–207
22. M. Prokopenko, F. Boschietti, A.J. Ryan, An information-theoretic primer on complexity, self-organization, and emergence. Complex. **15**(1), 11–28 (2009)
23. A.S. Ribeiro, S.A. Kauffman, J. Lloyd-Price, B. Samuelsson, J.E.S. Socolar, Mutual information in random Boolean models of regulatorynetworks. Phys. Rev. E **77**(1), 011901 (2008)
24. C. Gershenson, Phase transitions in random Boolean networks with different updating schemes (2004), arXiv:nlin/0311008v1. url:http://arxiv.org/abs/nlin/0311008

Chapter 8
Information Transfer in Biological and Bio-Inspired Systems

The main thrust of the preceding chapters has been the presentation of a framework for the local information dynamics of computation. Importantly, we have demonstrated that the underlying measures align with our qualitative notions of information storage, transfer and modification in analysing well-understood *theoretical* systems, including CAs and RBNs. We have also demonstrated that the perspective of local measures, i.e. quantifying the information measures at each point in space and time, produces more detailed insights into a given computation than averaged measures can.

In this chapter then, we will demonstrate the utility of the local perspective of information transfer in a number of *biological* and *bio-inspired* applications.

We will show that this local perspective can produce new insights in *understanding* interaction in biological systems. In Sect. 8.1 we show that examining local information transfer *in time* produces more detailed insights than averages into the time dynamics of the complex interaction between heart and breath rate in sleep apnea. Then in Sect. 8.2 we demonstrate that examining local information transfer *in space* is useful in the domain of computational neuroscience, by showing that it can establish the directed structure between brain regions that underpins a visuo-motor tracking task. The information transfer-based method we present is novel in the domain of computational neuroscience in providing directional, non-linear, model-free, collective analysis at the regional level, and being applicable to small data sets.

From a *design* perspective then, we will demonstrate in Sect. 8.3 that maximising local information transfer is a useful intrinsic goal in guided self-organisation for a bio-inspired snake-like robot (*snakebot*). We also show that local transfer entropy is unique in revealing the emergence of coherent information structure (akin to gliders) in this snakebot.

In all of these applications, the local perspective is crucial in providing full insights into the distributed computation of the system. The promising results here suggest further scope for the application of information dynamics in the realms of

J. T. Lizier, *The Local Information Dynamics of Distributed Computation in Complex Systems*, Springer Theses, DOI: 10.1007/978-3-642-32952-4_8,

© Springer-Verlag Berlin Heidelberg 2013

computational neuroscience, guided self-organisation, and other biological and bio-inspired domains.

8.1 Heart and Breath Rate Interaction in Sleep Apnea

Sleep apnea (or apnoea) is a disorder characterised by pauses and bursts of breathing during sleep. These episodes, each known as apnea, seem to occur when the heart rate breaches some threshold [1]. The interaction between heart and breath rate in a sleep apnea patient was studied from the perspective of information transfer between the two by Schreiber [1] in the original presentation of the transfer entropy (TE). That study found information transfer appears to occur in both directions between heart and breath rate. This indicated a complex interaction, albeit one with greater transfer from heart to breath in alignment with the observation of apnea occurring when the heart rate crosses some threshold [1].

In this section we study the interaction using *local* transfer entropy (see Sect. 4.1.2). We demonstrate the local measure to deliver *more detailed insights* into the dynamics of the complex interaction between heart and breath rate in time than the averages do. The application to this particular example is pertinent because of its use in the original presentation of the TE.

The local transfer entropies were computed in each direction between instantaneous heart and breath rate of a sleeping sufferer of sleep apnea (using samples 2350-3550 of data set B of the Santa Fe Institute time series contest held in 1991 [2], as per [1]). We use kernel estimation to compute the probability distribution functions of the relevant variables (see Sect. 2.2.3), with a kernel width of $r = 0.60$ standard deviations. Also, we use a history length of $k = 4$ (increasing beyond the $k = 1$ used in [1] to eliminate information storage as recommended in Chap. 4).

Figure 8.1c shows that significant bidirectional information exchanges between heart and breath rate coincide with the apnea events in the raw data of Fig. 8.1a and b (where the heart rate rises significantly and breath rate becomes highly variable). Significant clusters of spikes in the TE in both directions occur around these events, with little transfer taking place in between the events.

More importantly, the close-up of a typical event in Fig. 8.1d shows that **the information exchange is started by a significant transfer from heart rate to breath rate, followed by a complex two-way exchange of information during the apnea event**. This precedence in time is perhaps more indicative of dominance in the dynamics than having a larger average information transfer alone. This leading transfer from heart to breath is typical, but does not occur in all events; importantly, **it is the *local* transfer entropy that reveals the true richness of both the initial dynamics and the complex process that follows it**.

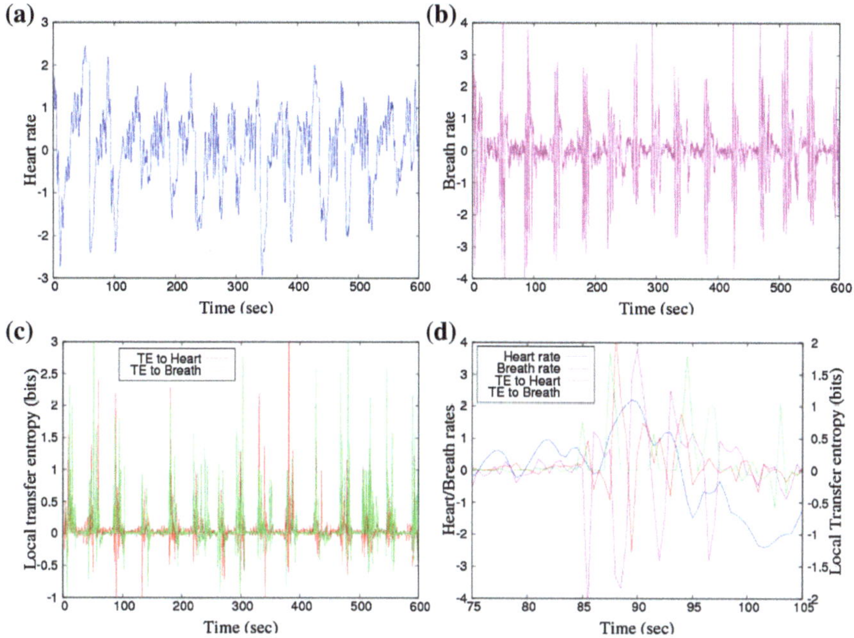

Fig. 8.1 Dynamics between heart and breath rates. **a** Heart rate, **b** breath rate: raw data normalised to zero mean and unit variance. **c** Transfers: local information transfer in each direction. **d** example of dynamics: raw data and local information transfers for an example apnea event

8.2 Establishing Directed Interregional Cortical Information Structure

The human brain undertakes highly sophisticated information processing facilitated by the interaction between its sub-regions. We present a novel effective interregional connectivity analysis of multivariate data. This analysis uses extensions to the transfer entropy, which is studied locally in space. The method allows us to identify the underlying directed information structure between brain regions, and how that structure changes according to behavioural conditions. This method is distinguished in using asymmetric, multivariate, information-theoretic analysis, which captures not only directional and non-linear relationships, but also the relationships due to collective interactions. We apply the method to analyse blood oxygen level-dependent (BOLD) time series to establish the directed information structure between brain regions involved in a visuomotor tracking task. Importantly, this results in a tiered structure, with known movement planning regions driving visual and motor control regions. Also, we examine the changes in this structure as the difficulty of the visuomotor tracking task is increased. We find greater coupling between sub-regions of a network involved in movement planning and areas involved in hand- and eye-movements (left SMA and left PMd, right cerebellum and right SC) with task difficulty. It is likely these methods will find utility in identifying interregional structure (and changes with experimental conditions) in other cognitive tasks and data modalities.

8.2.1 Introduction

Distributed computation in the brain is a complex process, involving interactions between many regions in order to achieve a particular task. It is particularly important for the different brain regions to interact and "integrate the disparate aspects of a cognitive process into a perceptual whole" [3]. Along these lines, Stevenson et al. have studied the areas where integration of audio-visual and visuo-haptic information occurs [4].

Our interest lies in establishing directed, interregional, effective[1] information structure in the brain based on particular cognitive tasks. Several studies in computational neuroscience have considered similar goals. By observing common information in different brain regions at different times, a flow of information was inferred between different regions in [7, 8]. The (pairwise) Granger causality was used in [9] for analysing interregional connectivity in a visual attention task. Effective networks were inferred with a measure of information transfer in a model of the macaque cortex in [6]. Also, single-variate information transfer was studied at the spiking neuron level in [10] and at the regional level in fMRI measurements in the human visual cortex in [11]. Undirected structure was studied with multivariate analysis at the regional level in fMRI measurements during visual processing in [12]. Other information-theoretical measures were used to study transfer between brain areas of macaques in [13].

In establishing such information structure, we are particularly interested in capturing *non-linear*, *directional*, *collective* interactions between different areas of one region facilitating outcomes in another region. Some of the above studies use linear methods only (e.g. Granger causality in [9], and linear approximations in [13]), others do not capture where an effect is due to a collective interaction[2] from two or more driving elements (e.g. [11] examines only average values over all variables in a region), or do not specifically examine the regional level of interactions. Also, some are not direct measures of information transfer, rather inferences of (undirected) common information (e.g. [12]).

---

[1] Note that the interregional information networks inferred here are effective networks rather than structural networks [5, 6]. Effective networks consist of a set of directed links representing statistical dependencies between the nodes in an underlying structural network. Effective networks provide insight into the *logical* structure of the network and how this changes as a function of network activity (regardless of whether the underlying structure is known).

[2] As described in Sect. 4.1.3, a collective interaction occurs for example in the XOR operation, where the outcome is not due to the isolated actions of single sources. Collective interactions are captured as interaction-based transfer in the complete and collective TE.

Here, we present a new approach in Sect. 8.2.2 to detecting directed information structure between brain regions in cognitive tasks.[3] Our approach examines the **statistical significance** of an ensemble of information transfer measurements between each region pair. These information transfer measurements are local in space, and each examines multiple variables (volume pixels or *voxels* for fMRI) in the source and destination brain regions. It is distinguished in using **asymmetric**, **multivariate**, **information-theoretical**[4] analysis, which captures not only **non-linear** relationships, but also **collective interactions** arising from groups of up to seven voxels in each region, and the **direction** of these relationships. Of particular importance is that this approach considers the collective interactions resulting from the combined activity of multiple voxels in each region: multi-voxel analysis is necessary in order to detect complete spatiotemporal patterns of activity [18], which are known to be more informative about experimental conditions than single voxel activity. The approach is further distinguished in using information theory to directly measure asymmetric information transfer, and this also makes it **model-free**. Also, the particular combination of information theory and multi-voxel analysis here is novel, as is the ability of the technique to provide insights from **relatively small data sets** due to our focus on statistical significance and use of innovative techniques for estimating the PDFs of continuous variables. The approach can also be used to infer differences in the structure as the cognitive task changes.

We apply our method in Sect. 8.2.3 to study fMRI[5] recordings of subjects undertaking a visuomotor tracking task, under various task difficulties. We note the important relationship between studying the structure of the perception-action loop in the human brain and studies in the domain of artificial intelligence, e.g. [19]. Our analysis of the fMRI data here yields a distinct, tiered, directed interaction structure which connects movement planning regions (as information sources) to visual and motor control regions (as information destinations). The correlations of the strength of the interregional relationships with task difficulty are then analysed to determine which pairs of regions have: a. more in common, or b. a more pronounced directional relationship as the task difficulty increases. Most significantly, we identify an increased coupling between regions involved in movement planning (left SMA and left PMd) and execution (right cerebellum for right hand and right SC for eye movements) with task difficulty. The method is thus demonstrated to be useful for investigating interregional structure in fMRI studies, but it can certainly

[3] The algorithm for inferring directed interregional information structure, and results of its application to fMRI data from a visuomotor tracking task were first reported in [14, 15].

[4] Information theory is known to produce useful insights from analysis of fMRI images, e.g. [16, 17]. The novelty here lies in its combination with asymmetric, multivariate and statistical significance-based techniques to infer directed information structure.

[5] Functional magnetic resonance imaging (fMRI) is a brain imaging technique which produces a multivariate time-series of measurements for many spatial points (*voxels*) within the brain at high spatial resolution (typically 3 mm^3 volumes). The measurements are of changes in blood flow and oxygenation related to neural activity near that point.

also be applied to other modalities such as electrophysiological multi-electrode array recordings.

8.2.2 *Interregional Information Structure Analysis Technique*

We begin by presenting our method for detecting directed information structure between brain regions in cognitive tasks. *Information theory* is the natural domain for measuring information transfer between brain regions, providing a model-free, non-linear platform for quantifying the information content of individual variables, variable collections or exchanges between variables.

The basic measures used here are the mutual information (MI, see Eq. (2.6)) for undirected measurements of statically shared information, and the transfer entropy (TE, see Eq. (4.1)) for directed measurements of dynamics information transfer. Here, the MI is estimated for continuous-valued variables using the techniques of Kraskov et al. [20, 21] (see Sect. 2.2.3), with a window size of the $K = 2$ closest observations. For the TE, we will use a history length $k = 1$ due to limitations of the number of observations. While the TE is not a direct measure of causal effect, the use of this short history length alters the character of the measure towards inferring causal effect (see Sect. 4.4.3). The TE is computed here as two MI terms (see Eq. (4.2)) using MI estimators for each in the style of [20, 21].[6] Again, we use a window size of the $K = 2$ closest observations for the Kraskov-estimators here. The use of Kraskov-estimators is important since they were specifically designed to handle small numbers of observations.

We have already shown in Chap. 4 that the TE is a non-linear, directional measure for the information transfer between two variables. In the domain of computational neuroscience, it has been used for example to analyse cortical interactions in simulated data [6], EEG data [24, 25], and fMRI data [10, 11], while a related measure (the "directed transinformation") was used in the investigations in [13]. As we have already pointed out in Sect. 4.1.3, the (apparent) TE does not detect interaction-based transfer due to two or more source variables (e.g. an XOR operation). To capture this aspect, we need to extend it to consider *multivariate* sources and destinations.

We outline how to extend these basic quantities to measure multivariate and interregional information transfer in Sects. 8.2.2.1 and 8.2.2.2. Subsequently, we describe how to assess the statistical significance of the interregional information transfer in order to infer directed information links from these measurements in Sect. 8.2.2.3. Finally, we describe how to detect changes in the directed structure as the cognitive task changes in Sect. 8.2.2.5.

---

[6] Note the TE could be computed using Kraskov estimation [20, 21] but with a direct conditional MI calculation as per [22, 23].

8.2.2.1 Extending Information Transfer to Multivariate Source and Destination

Each of these measures of information transfer may be trivially extended to consider joint variables as the source and destination, i.e. we have $I(\mathbf{X}; \mathbf{Y})$ and $T_k(\mathbf{Y} \rightarrow \mathbf{X}) = I(\mathbf{Y}; \mathbf{X}' \mid \mathbf{X}^{(k)})$ where $\mathbf{X}$ and $\mathbf{Y}$ are joint variables.[7]

The **multivariate mutual information** $I(\mathbf{X}; \mathbf{Y})$ measures the amount of information shared between a set of source variables $\mathbf{Y}$ and a set of destination variables $\mathbf{X}$. It has been studied for example in fMRI data in [12].

The **multivariate transfer entropy** $T_k(\mathbf{Y} \rightarrow \mathbf{X})$ (see Fig. 8.2) measures the amount of information that a set of source variables $\mathbf{Y}$ provide about a set of destination variables $\mathbf{X}$, that was not contained in the past of the destination set.[8] While the multivariate TE has been considered elsewhere (e.g. in genetic microarray data sets [26][9]), we are not aware of its application to fMRI data as yet. This is an important extension in this context, because multivariate voxel patterns are known to be more informative about experimental conditions than single voxel activity [27]. In particular, $T_k(\mathbf{Y} \rightarrow \mathbf{X})$ will capture both single-source and interaction-based transfer; the same cannot be said for single voxel analysis. As such, in the multivariate TE we now have a measure for information transfer that is non-linear, directional and captures collective interactions.

Note however that the number of available observations limits the number of joint variables that may be analysed in this manner (e.g. see [20, 28]), in a similar fashion to the history length k being limited for the active information storage and TE (see Sects. 3.3.1 and 4.1.1.1).

8.2.2.2 Measuring Information Transfer Between Two Regions of Variables

Though theoretically appealing, it is generally impractical for us to compute the **interregional information transfer** as the multivariate TE $T_k(\mathbf{R}_a \rightarrow \mathbf{R}_b)$ between the complete sets of variables in two regions $\mathbf{R}_a$ and $\mathbf{R}_b$. This is because a "region" in neural applications can contain a very large number of variables,[10] and complete sampling of these massively multivariate spaces would require many orders of magnitude more observations in time than could be practically obtained. As such, the

---

[7] We alter our information-theoretical notation in this section from $I_{X;Y}$ to $I(X; Y)$ in order to accommodate the complex notation for the variables under consideration.

[8] Note the similarity between the multivariate TE and the collective TE defined in Eq. (4.41). The distinction is that here we have a multivariate destination, whereas the collective measure considers a multivariate source only. This extension eliminates information that was already present in any of the destination variables. This is important for considering regions of variables, where one wishes to eliminate information already contained elsewhere in the region (even if that information moves around the variables within that region). The collective TE focuses on the information added by multiple sources to a single destination only.

[9] More specifically, this appears to be a consideration of the collective TE.

[10] For example, fMRI regions contain potentially hundreds of voxels, see Table 8.1.

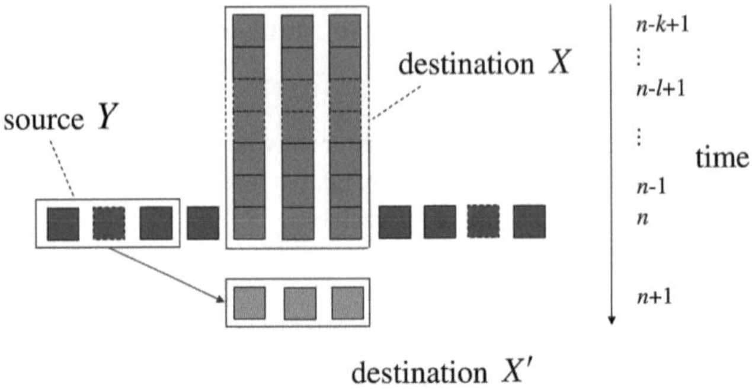

Fig. 8.2 Multivariate transfer entropy $T_k(\mathbf{Y} \to \mathbf{X}) = I(\mathbf{Y}; \mathbf{X}' \mid \mathbf{X}^{(k)})$ from the set of source variables $\mathbf{Y}$ to the set of destination variables $\mathbf{X}$. (NB: This figure is reprinted from [15] with permission of Springer)

most practical way to retain the benefits of this measure is to compute the interregional information transfer from region $\mathbf{R}_a$ to region $\mathbf{R}_b$ as the multivariate transfer entropy *averaged* over a large number S of sample pairs of subsets of v variables in each region:

$$T_{k,v}(\mathbf{R}_a \to \mathbf{R}_b) = \left\langle T_k(\mathbf{R}_{a,i} \to \mathbf{R}_{b,j})\right\rangle_{i,j}. \tag{8.1}$$

Here i and j label the subsets $\mathbf{R}_{a,i}$ and $\mathbf{R}_{b,j}$ of size v in each region. Again, while it is desirable to average over all pairs of subsets from each region, where:

$$S = \binom{|\mathbf{R}_a|}{v}\binom{|\mathbf{R}_b|}{v}. \tag{8.2}$$

this is impractical for large regions. In practical situations, S must be limited with the subsets selected randomly. Thus, $T_{k,v}(\mathbf{R}_a \to \mathbf{R}_b)$ provides a measure of interregional information transfer that is non-linear, directional, captures collective interactions and is practically usable.

The same technique of averaging over subsets of size v can be used with the multivariate MI to produce a similar but non-directional measure, the **interregional mutual information**:

$$I_v(\mathbf{R}_a; \mathbf{R}_b) = \left\langle I(\mathbf{R}_{a,i}; \mathbf{R}_{b,j})\right\rangle_{i,j}. \tag{8.3}$$

Indeed, this is done in [12], where the authors additionally condition on experimental conditions.

Finally, we note that both measures may be averaged over subjects s in order to assess group effect. This gives the following expressions:

$$T^g_{k,v}(\mathbf{R}_a \to \mathbf{R}_b) = \left\langle T_{k,v}(\mathbf{R}_a \to \mathbf{R}_b)\right\rangle_s, \tag{8.4}$$

$$I^g_v(\mathbf{R}_a; \mathbf{R}_b) = \langle I_v(\mathbf{R}_a; \mathbf{R}_b)\rangle_s. \tag{8.5}$$

8.2.2.3 Significance Testing of Information Transfer Measures

The measures described above quantify the information transfer between single or sets of variables. However, it is important to realise that since they are computed from a finite number of observations they are actually *random variables*. As a consequence, even where there may be no temporal relationship between two variables, there is a non-zero distribution of results for a transfer entropy measurement between them based on a finite number of observations. Another consequence is that even with a strong relationship between some subsets of variables in each region, $T_{k,v}(\mathbf{R}_a \rightarrow \mathbf{R}_b)$ may average over many other pairs of subsets that have no temporal relationship to each other, resulting in what may appear to be a low average value.

The intersection of these factors means that we need an objective method of determining whether a value of $T_{k,v}(\mathbf{R}_a \rightarrow \mathbf{R}_b)$ indicates a *significant* directional relationship $\mathbf{R}_a \rightarrow \mathbf{R}_b$. Simple thresholding is undesirable since the choice of threshold is subjective and may not be comparable between different region pairs. Considering the **statistical significance** of the measurement is the most desirable approach: i.e. determining the probability that the measured value of $T_{k,v}(\mathbf{R}_a \rightarrow \mathbf{R}_b)$ or greater would have been observed were there no temporal relationship between the source and destination regions. In this section we describe how to determine the statistical significance of $T_{k,v}(\mathbf{R}_a \rightarrow \mathbf{R}_b)$ in order to infer directed interregional links.

First, we consider statistical significance testing for the single-variate TE measurement $T_k(Y \rightarrow X)$, as previously described in [24, 29]. The *null hypothesis* H_0 of the test is that the state changes $x_n^{(k)} \rightarrow x_{n+1}$ of the destination X have no temporal dependence on the source Y.

Assuming H_0 true, we need to determine the distribution of TE measurements $T_k(Y^p \rightarrow X)$ under this condition. This is done[11] by generating many surrogate time series (say P of them) Y^p by randomly permuting the source time series Y, then using each one to compute a surrogate TE to X: $T_k(Y^p \rightarrow X)$. Importantly, these surrogates are computed from the same number of observations, and the same distributions $p(y_n)$ and $p(x_{n+1} \mid x_n^{(k)})$; the only difference is that the temporal dependence $p(x_{n+1} \mid x_n^{(k)}, y_n)$ of the state changes of the destination on the source has been destroyed. Thus the distribution of the surrogates $T_k(Y^p \rightarrow X)$ describes our expectation for $T_k(Y \rightarrow X)$ under H_0. We can then determine a p-value as the probability $p\left(T_k(Y^p \rightarrow X) \geq T_k(Y \rightarrow X)\right)$ using Student's t-test; i.e. the probability of observing a greater $T_k(Y \rightarrow X)$ than that actually measured, assuming H_0.

For a given α value, we reject H_0 when $p < \alpha$, concluding then that a significant temporal relationship between the source and destination does exist. Significance of the MI may be determined in a similar fashion. This method provides an objectively determined threshold for any TE measurement, and the given α allows comparison to other pairs of variables.

[11] The following explanation assumes that only one previous state y_n of the source is used in the computation of $T_k(Y \rightarrow X)$; i.e. the parameter $l = 1$ (see Eq. (4.1)).

8.2.2.4 Significance Testing of Interregional Transfer Entropy

Significance testing of the multivariate measures $I(\mathbf{X};\mathbf{Y})$ and $T_k(\mathbf{Y} \rightarrow \mathbf{X})$ is a straightforward extension. Most importantly, in generating the surrogates $\mathbf{Y}^p$ we do not permute the component time series $Y_1, Y_2, \ldots$ of $\mathbf{Y}$ *individually* but permute the vectors $\mathbf{y}_n$ at each time point n in $\mathbf{Y}$ *as a whole*. This ensures that the only difference in making the surrogate measurements $T_k(\mathbf{Y}^p \rightarrow \mathbf{X})$ is the temporal relationship $p(\mathbf{x}_{n+1} \mid \mathbf{x}_n^{(k)}, \mathbf{y}_n)$.

Similarly, significance testing can be extended to the interregional measures $T_{k,v}(\mathbf{R}_a \rightarrow \mathbf{R}_b)$ and $I_v(\mathbf{R}_a;\mathbf{R}_b)$ by generating P surrogate measurements as, for example:

$$T_{k,v}(\mathbf{R}_a^p \rightarrow \mathbf{R}_b) = \left\langle T_k(\mathbf{R}_{a,i}^p \rightarrow \mathbf{R}_{b,j}) \right\rangle_{i,j}. \tag{8.6}$$

Note that the pth permutation is applied to the whole region $\mathbf{R}_a^p$ *before* the subsets i of v variables are selected for $\mathbf{R}_{a,i}^p$ from it. Also, for each of the P permutations $\mathbf{R}_a^p$ the same subsets i and j must be selected as for the actual measurement $T_{k,v}(\mathbf{R}_a \rightarrow \mathbf{R}_b)$ in Eq. (8.1). Together, these constraints ensure that the only difference in making the surrogate measurements $T_{k,v}(\mathbf{R}_a^p \rightarrow \mathbf{R}_b)$ is the temporal relationship $p(\mathbf{r}_{a,n+1} \mid \mathbf{r}_{a,n}^{(k)}, \mathbf{r}_{b,n})$. As such, we can reject H_0 and conclude a significant interregional link if[12]:

$$p\left(T_{k,v}(\mathbf{R}_a^p \rightarrow \mathbf{R}_b) \geq T_{k,v}(\mathbf{R}_a \rightarrow \mathbf{R}_b)\right) < \alpha. \tag{8.7}$$

This method of identifying significant interregional links is suitable for application on the *individual level* (e.g. for fMRI measurements from an individual subject). We are particularly interested in extending this inference to the *group level* where we have measurements for each region for a number of subjects within a group. To do so we need to compare the measured averages across the group (i.e. $T_{k,v}^g(\mathbf{R}_a \rightarrow \mathbf{R}_b)$ from Eq. (8.4)) to the distribution of P group averages obtained from the surrogate measurements at each subject, e.g.:

$$T_{k,v}^g(\mathbf{R}_a^p \rightarrow \mathbf{R}_b) = \left\langle T_{k,v}(\mathbf{R}_a^p \rightarrow \mathbf{R}_b) \right\rangle_s. \tag{8.8}$$

We can reject H_0 and conclude a significant interregional link at the group level if:

$$p\left(T_{k,v}^g(\mathbf{R}_a^p \rightarrow \mathbf{R}_b) \geq T_{k,v}^g(\mathbf{R}_a \rightarrow \mathbf{R}_b)\right) < \alpha. \tag{8.9}$$

We now have a method to determine which region pairs have a significant directed relationship under a given cognitive task at the group level. The set of these directed relationships form a directed interregional information structure.

---

[12] We note the different approach taken to inferring links with the interregional MI in [12], where the authors examined its statistical significance as compared to the average over multivariate MIs computed from random subsets of v variables taken from any of the brain regions.

8.2.2.5 Changes in Structure with Task Conditions

Finally, we consider how to identify statistically significant changes in the interregional structure as the experimental conditions change for a given cognitive task. At the *group level*, for a given region pair we can compare the populations of interregional measures $T_{k,v}(\mathbf{R}_a \rightarrow \mathbf{R}_b)$ and $I_v(\mathbf{R}_a; \mathbf{R}_b)$ for all subjects s between the different experimental conditions. If there are two experimental conditions we compare the populations using a t-test with the null hypothesis that there is no difference between the populations. If the experimental condition is a quantity (e.g. a task difficulty level as per the task we analyse in Sect. 8.2.3) then we test the population of correlation scores (of interregional measure to task difficulty) for each subject against the null hypothesis of a zero correlation.

With the TE measurements, these tests determine changes in *directed* information transfer for a given region pair as a function of experimental condition. With the MI measurements, the tests determine whether the regions have more or less *in common* as as a function of the experimental condition.

8.2.3 Application to fMRI Experimental Data

In this section, we describe the application of the above methods for determining interregional structure to brain imaging data from a visuomotor tracking task. We also examine the changes in this structure as the difficulty of the task is altered.

8.2.3.1 Visual Tracking Task and BOLD Acquisition

The cognitive task for this experiment involved eight subjects (who gave informed written consent) tracking a visual target moving along a circle on a computer screen with their right index finger.[13] The task required the integration of information from visual input, movement planning and visual and motor control regions. The difficulty of the tracking task was altered between four different levels in blocks of 16.8 s, with 20 such blocks sampled for each difficulty.

During the cognitive task, functional Magnetic Resonance Imaging (fMRI) EPI data was acquired from each subject. With a 2.8 sec sampling interval, we have 140 observations in time for each difficulty level for each subject. The fMRI imaging had a spatial resolution of 3 mm^3. The data was pre-processed and extracted for 11 regions active during the task (see footnote 13). The 11 regions in the data set, their abbreviations and number of voxels (volume pixels) are listed in Table 8.1.

[13] The fMRI data set was obtained by researchers from the Bernstein Center for Computational Neuroscience in Berlin (see Acknowledgements on p. 11). Furthermore, the data set was subjected by these researchers to standard preprocessing operations (motion correction, spatial normalisation), and a general linear model (GLM) was used to find regions that were activated during the task. The data collection and these prior analyses are described in [30].

Table 8.1 Table of brain regions analysed in the visuomotor tracking task

Name of region	Abbreviation	Number of voxels
Left superior colliculus	Left_SC	42
Right superior colliculus	Right_SC	35
Right cerebellum	Right_Cerebellum	185
Right basal ganglia	Right_BG	74
Primary visual cortex	V1	63
Left primary motor cortex	Left_M1	158
Left supplementary motor area	Left_SMA	24
Left dorsal premotor cortex	Left_PMd	296
Right dorsal premotor cortex	Right_PMd	32
Left superior parietal lobule	Left_SPL	56
Right superior parietal lobule	Right_SPL	43

8.2.3.2 Directed Information Structure

For each region pair $\mathbf{R}_a$ and $\mathbf{R}_b$ here, $T_{k,v}(\mathbf{R}_a \rightarrow \mathbf{R}_b)$ and $I_v(\mathbf{R}_a; \mathbf{R}_b)$ were measured for all subjects s using $v = 3$ and 5 voxels for transfer entropy and $v = 5$ and 7 voxels for mutual information. We average over at least $S = 1000$ subset pairs for each interregional information transfer measurement, with the subsets selected randomly. The averages across the groups $T^g_{k,v}(\mathbf{R}_a \rightarrow \mathbf{R}_b)$ and $I^g_v(\mathbf{R}_a; \mathbf{R}_b)$ were significance tested as described in Sect. 8.2.2.4 in order to determine an interregional information transfer structure. We used at least $P = 100$ surrogate time series for the significance tests (with $\alpha = 0.05$) as described in Sect. 8.2.2.4. It is important to note that the method does not exclude bidirectional relationships (i.e. it does not just look for which direction was strongest). The structures were additionally filtered by requiring interregional links to qualify as significant at both subset sizes v tested for each measure, and for all four task difficulty levels; such additional filtering is a strong precaution against false discovery rate.

Significance testing of the interregional MI detects an almost fully connected undirected structure (115 out of 120 undirected region pairs connected). In one respect this is insightful, as it indicates that all of the regions have much in common as the subjects undertake this cognitive task. However, it does not help to determine the directed information structure in place here.

On the other hand, significance testing of **the interregional transfer entropy produces a distinct three-tier directed information structure** (see Fig. 8.3).[14] This result is very informative about the information structure during the cognitive task. At the top tier are movement planning regions: left and right SPL, left and right PMD, left SMA and left primary motor cortex. These regions provide directed inputs to the middle and bottom tiers. The middle tier contains (visual) sensor processing and control regions: left and right SC and primary visual cortex. While receiving input from the top tier, the middle tier also provides directed input to the bottom tier.

[14] Figures 8.3 and 8.4 were generated using Cytoscape [31].

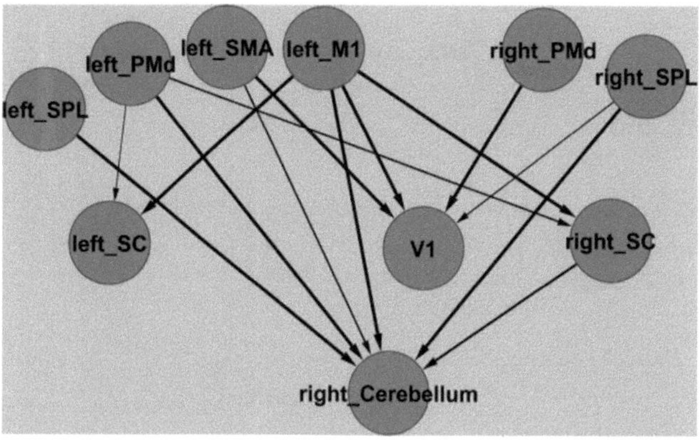

Fig. 8.3 Directed information structure established with the interregional transfer entropy for the visuomotor task. The *top* tier represents includes planning regions, the *middle* tier (visual) sensor processing and control regions, and the *bottom* tier motor execution. *Thickness of lines* indicates significance against the null hypothesis of having no temporal relationship within a pair (thickest: $p < 0.01$, thinnest: $p < 0.05$). (NB: This figure was first published in [14])

The bottom tier contains motor execution in the right cerebellum. Note that the BG region is not found to be involved in the directed structure.

This tiered structure correlates well with the experiment, where one would expect movement planning areas to direct the eyes to track the object (via SC regions) and provide information on this movement to be integrated with visual sensing, as well as to direct the motor execution for tracking the object with the right index finger (via the right cerebellum). Additionally, the visual processing of the object would be expected to be predictive of the motor execution to track the object.

One could also expect to find some level of feedback from the visual sensing regions to the movement planning (since planning should require information of where the object is), however such feedback links are noticeably absent from the results in Fig. 8.3. This is possibly because such feedback links operate on a shorter time-scale than the fMRI measurements here (at >2 second intervals). That is to say, the raw visual information about where the object is may produce a detectable influence on the activity in the movement planning regions for only a short time frame (i.e. within two seconds and no longer). Comparing to the directed links that are detected, it is reasonable to explain that movement planning regions may have predictive effects on the visual and motor control regions over fMRI time scales via high level concepts such as direction and speed.

8.2.3.3 Changes in Directed Information Structure

Significance testing of the correlation of interregional information transfer with task difficulty was performed for each region pair, using both TE and MI (to assess the

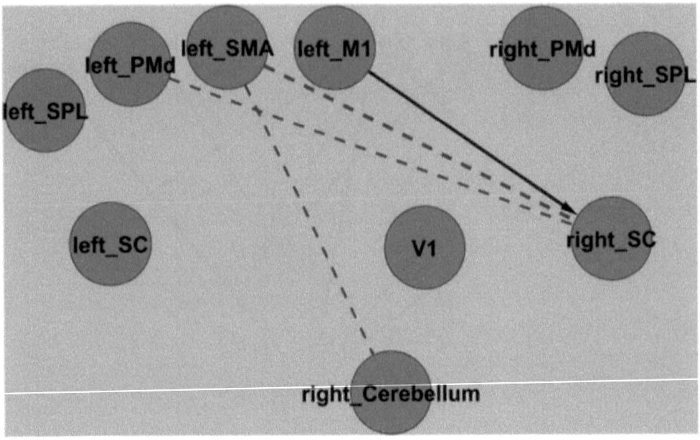

Fig. 8.4 Connections in the directed information structure with a statistically significant (t-test, $p < 0.05$) correlation between information (transfer) and difficulty of the visuomotor task. Key: *dashed green line* = correlation with interregional mutual information; directed *red line* = an anti-correlation with the directed interregional transfer entropy. (NB: This figure was first published in [14])

changing nature of the directed and undirected relationships respectively) as outlined in Sect. 8.2.2.5. Here, we use one-sided t-tests with $\alpha = 0.05$. We also add the additional filtering requirement (to combat false discovery rate) that significant correlations must be detected for the pair for both subset sizes tested for the given measure. Furthermore, we require that the pair must have also been concluded to have a significant link (with the given measure) in the previous section to ensure the correlation was not on spurious values. For example, we only draw conclusions about significant changes in TE for links established in Fig. 8.3.

These tests reveal few changes in the information structure as the task difficulty is increased. As shown in Fig. 8.4, we find one directed region pair which has a significant decrease in interregional TE with task difficulty, and three region pairs with a significant increase in interregional MI with task difficulty. These changes are interesting because they **indicate an increased coupling between regions involved in movement planning** (left SMA and left PMd) **and execution** (right cerebellum for hand movements, right SC for eye movements) **as the task difficulty increases**. Though this result does not indicate an increase in directional transfer between these areas, the increase in coupling between them is indeed expected.

It is possible that there was a real though undetected increase in directional information transfer with task difficulty on many of the links established in Fig. 8.3. The increased information transfer may have been at shorter time scales than fMRI sampling rates allow us to detect. It is also possible that there is in fact no increase in directional information transfer in any of these channels, simply a change in what the information represents. As sampling rates improve, future experiments will be able to investigate these possibilities further.

8.2.4 Conclusion

We have presented a novel approach to analysing effective connectivity of time-series data to establish interregional information structure. Its combined characteristics (being information-theoretical, asymmetric and multivariate) distinguish it in identifying directional, non-linear, and collective interactions between regions in a model-free manner. It is also applicable to small numbers of observations due to our use of Kraškov-estimators coupled with our statistical significance perspective.

Here, the method was applied to fMRI data from a visuomotor tracking task. The 140 time steps available in each fMRI series would typically be considered too short for multivariate information-theoretic analysis; however the aforementioned features of our technique allow it to provide insights here. It identified an interesting three-tier interregional information structure, with movement planning regions providing input to visual perception and control regions, and both these tiers driving motor execution. The method also identified increased coupling between movement planning and motor execution regions as the tracking task became more difficult.

The presented method has thus been demonstrated to be useful for investigating interregional structure in fMRI studies, but it could also be applied to other modalities such as electrophysiological multi-electrode array recordings.

Future work will involve investigations of information transfer on shorter time scales as measurement technology improves. Also, while we have demonstrated the utility of studying information transfer on a *spatially* local scale in the domain of computational neuroscience, we would like to explore information transfer in the cortex on a local scale in *time* as well as space. This may result in the identification of coherent information transfer structures (akin to gliders in CAs) in the cortex, as described in [32].

8.3 Evolution of Coherent Information Transfer Structure

As discussed in Sect. 2.5, several authors have recently been investigating the potential for *guided self-organisation* in system design, e.g. [33–37]. This concept proposes the use of information-theoretical measures of the information processing carried out by a system in order to guide its evolution. An initial approach is to use these measures as generic fitness functions in evolutionary design. From an engineering perspective, template-based methods for generic information processing skills could be simpler and afford a framework-based approach to such design of self-organised systems. It also provides to us the potential to better understand the evolved solutions, and more importantly the opportunity to study and understand the emergence rather than engineering of intelligence [34].

We believe that a promising approach to guided self-organisation is a focus on the use of measures of the information dynamics of distributed computation (from Chaps. 3–5). Any task we wish to evolve the system to solve involves a distributed computation, so evolving for the fundamental building blocks of the computation is a direct way to allow that computation to emerge. We could evolve directly for a particular computational property (e.g. information storage as opposed to transfer), or for a mix of those properties. As described in Sect. 2.5, information transfer has been suggested to be a particularly important fitness function here, primarily because of its important role in distributed computation.

In this section, we present the first use of a direct measure of information transfer, transfer entropy [1], as the sole fitness function in an evolutionary design task.[15] An initial aim of the experiment is to check whether information transfer underpins coordinated motion, as was suggested in previous work [33]. More importantly, we aim to investigate what type of behaviour emerges when a system is evolved to maximise information transfer. Much previous work on information-driven evolution has sought to confirm whether it can approximate direct evolution for a given task. Here, we simply seek to investigate what type of solution or computation is generated by evolution for information transfer, and hypothesise that it will induce useful computation in the system. Our findings will help us to understand the role that information transfer can play in a unified framework for guided self-organisation, focusing on the information dynamics of distributed computation.

We use a snake-like modular robot (the *snakebot*) for experimentation. Information structure has been observed to emerge previously in snakebots when using a fitness function for fastest motion [39], and conversely fast motion has emerged from evolution with a measure of coordination as the fitness function [33]. We measure information transfer using the apparent transfer entropy (see Sect. 4.1.3) between neighbouring modules of the snakebot, and evolve the snakebot to maximise this quantity. This information transfer could be utilised by the snake in leading to coordinated motion between the modules, communicating information about obstacles, or driving new behaviours in a given direction along the snake.

We report that *coherent* information transfer structures were observed to emerge (using local transfer entropy, see Chap. 4) in the evolved snakebot. We say "emerged" because while high information transfer was selected for, local *coherent* structures were not part of the specification.[16] This is an important finding, because these structures are analogous to *glider* structures in cellular automata (CAs). Gliders are known to be the information transfer agents in CAs, providing for long-range correlations across space and time and playing a fundamental role in the distributed computation carried out in the CA (see Chap. 4). As such, we have provided evidence that using a direct measure of information transfer as a fitness function in guided self-organisation can produce useful structure in the system.

[15] The application presented in this section was first reported in [38].

[16] As we will discuss later in this section, we saw in earlier chapters that the ECAs with the highest average information transfer values did not contain gliders. As such, selection for high transfer could not automatically be expected to result in glider-like coherent structures here.

Fig. 8.5 Snakebot (NB: This figure was first published in [38])

8.3.1 Evolving the Snakebot for Maximum Information Transfer

The snakebot is a snake-like modular robot, introduced by Tanev et al. [40], which is simulated in the Open Dynamics Engine (ODE).[17] As shown in Fig. 8.5, it consists of a set of identical spherical morphological segments which are linked by universal joints. The joints each have two actuators for joint rotation, which are oriented vertically and horizontally in the initial standstill position of the snakebot, and all have identical angle limits. No anisotropic friction between the morphological segments and the surface is considered. The genome for the snakebot is an algebraic expression for the *desired* turning angles of its horizontal and vertical actuators as a function of time and actuator index. The periodic functions sin and cos are included in the function set, providing support for periodic gaits. The turning angles however are constrained by interactions between the segments and with the terrain; as such the *actual* actuator angles represent the emergent dynamics. Here, $\alpha_{i,n}$ and $\beta_{i,n}$ represent the actual horizontal and vertical turning angles respectively at time step n, where i is the actuator index (so $1 \leq i \leq S$ where $S = 14$ is the number of joints), and $1 \leq n \leq N$ for $N = 1800$ time steps in the simulation run.

Initial experiments to evolve fastest motion in any direction indicated that side-winding motion (i.e. locomotion predominantly perpendicular to the long axis of the snakebot) provided superior speed characteristics [40]. As previously mentioned, subsequent experiments observed an increase in coordination (as excess entropy) with this evolution [39], and then evolved similar fast moving side-winding locomotion using this measure of coordination as a fitness function [33]. In capturing correlations across space and time, the (two-dimensional) collective excess entropy Sect. 3.1.1) is something of an overall measure of distributed computation which balances the underlying components of information storage and transfer between constituent variables. Here, we evolve the snakebot using transfer entropy, in order to maximise the information transfer component of distributed computation. It was suggested in [33] that information transfer underpinned coordinated motion. An

[17] The implementation of and original genetic programming framework for the snakebot were supplied by Ivan Tanev (see Acknowledgements on p. 11).

information transfer is certainly required in a transient sense to achieve coordinated motion, but the level of information transfer in this initial phase may not be very significant compared to the information transfer averaged over longer experimental periods for other behaviours. The evolution of the snakebot here will take place in a flat environment. We will observe what types of behaviour emerge as a result of selecting for information transfer.

In evaluating the fitness of each snakebot after it is simulated for N time steps, we compute the average apparent TE $T_{i+1 \rightarrow i}(k)$ between each pair of consecutive modules $i+1$ and i, in the direction from the tail toward the head (i.e. decreasing module number i). The TE is computed using the time series of actual horizontal turning angles $\alpha_{i,n}$. Kernel estimation is used with these continuous values (see Sect. 2.2.3), with resolution r set to one quarter of the standard deviation of the turning angles. Also, we use the default step kernel and maximum distance norm, ignoring matched pairs within 20 time steps *and* neighbouring modules to avoid spurious dynamic correlations (as recommended by Schreiber [1]). The direction of tail toward head is selected because each module only applies desired turning angles to the actuators in front of it (i.e. in the direction of the head), thereby giving preferential treatment to information travelling in this direction. Although it is possible for information to be transferred across more than one joint per time step, we consider only consecutive pairs since this is likely to be the dominant transfer mode. Also, as per Sect. 4.4.2, we only consider transfer from a single previous state of the source variable ($l = 1$ in Eq. (4.1)), so as to consider information transferred directly at the given time step. We use a past history length $k = 30$ (as for the correlation entropy calculations in [33]). This is large enough to eliminate information storage from the calculation (see Sect. 8.3.2), while allowing adequate sampling of the underlying distributions (because the presence of sin and cos functions mean that the emergent turning angle sequences are generally quasi-periodic and therefore much of the state space of $\alpha_{i,n}^{(k)}$ remains unexplored). Our *fitness function* is then the average of these transfer entropies over all $S-1$ consecutive module pairs for the given snakebot:

$$T_{tail \rightarrow head}(k) = \frac{1}{S-1} \sum_{i=1}^{S-1} T_{i+1 \rightarrow i}(k). \tag{8.10}$$

The Genetic Programming (GP) techniques used for snakebot evolution are described in [40]. The snakebots evolve within a population of 200 individuals, with the best performers selecting using the fitness function described above. No minimum limit is placed on how far the snakebot moves, since we are not evolving for fast locomotion. The selection is based on a binary tournament with selection ratio of 0.1 and reproduction ratio of 0.9. Random subtree mutation is used with a ratio of 0.01.

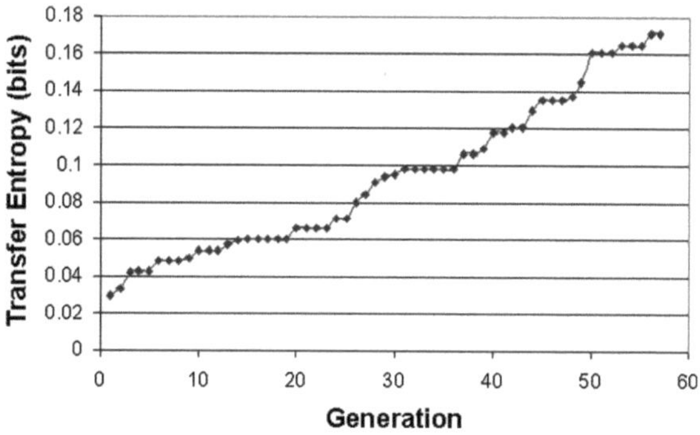

Fig. 8.6 Snakebot fitness (average transfer entropy $T_{tail \to head}(k = 30)$) per generation, plotted for the best performer in each generation. (NB: This figure was first published in [38])

8.3.2 Results and Discussion

First, we note that **snakebots exhibiting a high degree of coordinated motion** (as exemplified by the most fit individual from [33]) **were found to have significantly lower transfer entropy** than individuals specifically evolved to maximise transfer entropy (e.g. 0.007 bits versus 0.175 bits for the most fit snakebot here). Highly coordinated snakebots exhibited very short transients before becoming coordinated, and minimal TE in their ongoing behaviour. Coordinated motion is certainly more strongly associated with memory than information transfer. When neighbouring modules achieve perfect coordination, they have effectively reached a periodic attractor: their next states are completely predictable from their individual pasts, and so no additional information from the neighbour is measured as TE. This is analogous to the case of vanishing information transfer in the periodic background domains of CAs (see Sect. 4.2.3). It is possible that TE might be measured to be higher for snakebots attempting coordinated motion in a challenging environment, where information transfer in the longer and more significant transient toward coordination may play an important role in the dynamics.

In our evolution of snakebots for TE, the growth in the average TE $T_{tail \to head}$ $(k = 30)$ of the most fit snakebot in each generation is shown in Fig. 8.6.

We will focus on the most fit individual in the final (57th) generation as the result of this evolution, which had an average TE of 0.175 bits between neighbouring modules toward the head per time step. This snakebot did not display a fast, well coordinated side-winding locomotion. Instead, it displayed a complex form of wriggling behaviour, where thrashing of the tail appeared to drive new behaviour along the body of the snake, achieving a slow movement to the side.[18] The dynamics of this behaviour are more clear when examining the time-series of the actual horizontal turning angles $\alpha_{i,n}$, as displayed in Fig. 8.7a. Here, we see that coherent waves of behaviour are consistently travelling along the snakebot, from the tail toward the

[18] Videos of the snakebot, showing raw motion and local transfer entropy are available at http://lizier.me/joseph/publications/08ALifeSnakebotTe or http://www.prokopenko.net/modular_robotics.html or http://www.youtube.com/view_play_list?p=6604CF436CC0738C

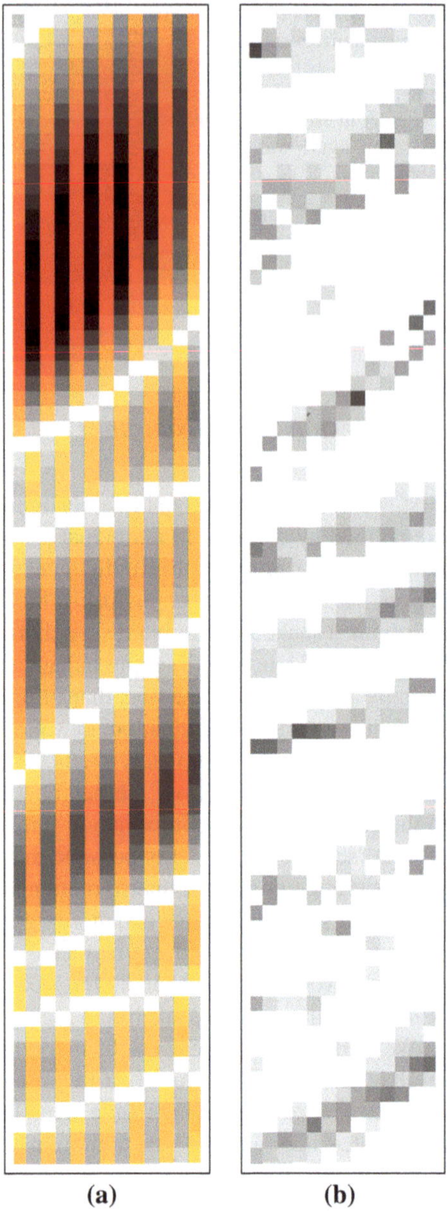

Fig. 8.7 Local apparent transfer entropy highlights coherent transfer structures or "gliders" in the evolved snakebot. **a** Raw actuator turning angles for each of the 13 destination modules (head at *left*, tail at *right*) of the snakebot for 76 consecutive time steps (time increases down the page): *greyscale* represents a positive turning angle, *yellow-red* represents a negative turning angle; range is −50 to 50°. **b** local transfer entropy $t_{i+1 \rightarrow i}(n, k = 30)$ into each of the 13 information destination modules of the snakebot, between consecutive modules in the tail → head direction: *greyscale*, range 0.0 bits (*white*) to 2.8 bits (*black*). Note that only positive local values are displayed here, and very few negative values were observed. (NB: This figure was first published in [38])

head. Each wave involves the modules turning in alternating directions along the snake, reaching a maximum angle then coming back to a rest position. The modules then swap their turning angles in the next wave. Importantly, these waves are not completely periodic, allowing scope for information transfer effects.

(a) (b) (c)

Fig. 8.8 Snakebot modules coloured to indicate incoming local transfer entropy (*black* is 0.0 bits, *red* is 2.8 bits) from the neighbouring module toward the tail, for three consecutive time steps. The information transfer from the tail appears to communicate a straightening behaviour here. (NB: This figure was first published in [38])

Already, we note a fairly clear correspondence to emergent travelling structures in microtubules and gliders in CAs, however to confirm the information transfer properties, we examine the local transfer entropy profile in Fig. 8.7b. **The local transfer entropy profile here tells us much more about the snakebot dynamics than either raw states or the average transfer entropy** (as was observed for CAs in Sect. 4.2 [41]). Importantly, this is the first investigation of the *local* TE values for direct measurement on continuous variables. As expected, we confirm that **we have coherent information structures moving along the snakebot from the tail toward the head**, which coincide in direction and approximately in time with the time-series waves previously observed. As an example, note the images of the snakebot in Fig. 8.8 with modules coloured to indicate local TE. Also, videos with the modules of the snake highlighted according to their local TE are available online (see footnote 18 on p. 195). We can be confident that the information transfer measured is not misattributed information storage, because our use of $k = 30$ considers a longer past history than the length of the time-series waves here. Note that these coherent transfer structures were not observed in fully-coordinated or random snakebots.

There is a wide variation in the types of such information transfer structures observed in Fig. 8.7b: some move faster than others (indicated by a flatter structure), some are more highly localised in time (thinner structures), some contain higher local transfer entropies (darker colouring), and some do not coherently travel the whole way along the body of the snakebot. Importantly, **none of these differences are detectable by superficial examination of the time-series of the actual actuator angles**. Indeed, apart from their coincidence in direction and approximately in time, there is little correspondence that is obvious to the observer between the time-series waves and the information structure. Certainly, there is no simple method of using the time-series waves to infer the location in time of the local information transfer structures: these are observed to begin and end at various time points within the time-series waves. Local TE reveals the *precise* space-time dynamics of the manner in which the tail drives new behaviour in the snakebot in a way not possible by examining the time-series alone.

These travelling coherent information structures are clearly analogous to gliders in CAs (e.g. see Fig. 3.4e). This finding is significant because of the important role that gliders play in CA dynamics, where they coherently transfer information in the collective computation of the CA. We previously noted that the coincidence of gliders and coherent information transfer with a maximisation of apparent TE in comparison to the complete measure (see Appendix G, in the context of Chap. 7). Here, we have demonstrated the emergence of glider-like structures when apparent TE is optimised, *without* explicitly selecting for such local coherence. At first glance, this suggests that coherent glider-like structures are perhaps the most efficient mode of apparent information transfer. More likely however,[19] **it suggests that glider-like structures may be an evolutionary attractor for the coherent communication of information between adjacent agents**. That is, glider-like structures perhaps require a minimum number of evolutionary steps, and/or perhaps there are many or easy-to-find evolutionary paths between these steps. It also suggests that this local optimisation (maximising transfer on each link via evolution) can lead to useful structure on a global or emergent scale. Perhaps an adaptive local strategy (as opposed to evolutionary) may lead to similar types of information structure.

The result is also significant because of the prevalence of glider-like structures in natural systems, e.g. dipole-dipole interactions in microtubules [42], or waves of directional change in schooling fish [43]; and in artificial evolution, e.g. the ϕ_{par} density classification rule [44, 45]. There are significant implications for these structures, which could have evolved to exploit this efficient mode of information transfer where coherent communication or effect over some distance is beneficial.

The coherence of glider structures is of particular importance to the computation in CAs; without coherence of information transfer, complex computation does not appear to take place (see Chap. 7). A second requirement for the implementation of arbitrarily complex, universal computation though must be *bidirectional* information transfer. Without this feature, these glider-like structures cannot interact and modify information. In this experiment, with strong information transfer encouraged in one direction only, although we have demonstrated the emergence of an important building block for non-trivial computation, we have evolved only a trivial type of computation.[20] In future work, we will build on our results here to evolve bidirectional information transfer for true distributed computation.

8.3.3 Conclusion

We have presented the first experiment of the use of transfer entropy as a generic fitness function in the domain of guided self-organisation. We have demonstrated that maximising information transfer in this manner can lead to the emergence of

---

[19] We know from the earlier chapters that the ECAs with the largest apparent TE values are not the ECAs exhibiting gliders, so it is unlikely that gliders are equivalent to a maximisation of TE at each separate communications link. We do know that ECAs with gliders exhibit the largest *proportion* of apparent to complete TE, so this proportion is perhaps related to communication over long distances rather than single communication links.

[20] This is effectively the reason that there are very few points of negative local TE measured in the snakebot here; see Fig. 8.7.

coherent information transfer structures which, as manifested by gliders, are known to underpin distributed computation in CAs. Importantly, the presence of these coherent structures could only be revealed by the *local* transfer entropy, in its first direct application to continuous variables here.

In this instance, the useful generic skill of coherent transfer structures was not fully capitalised on by the snakebot, but the important finding is that the use of information transfer as a fitness function led to the emergence of this computational capability. Also, our experiment implies that glider-like structures are the most efficient mode of coherent or single-source information transfer to evolve, which is itself significant insight into the nature of information transfer.

All agent-based systems compute; indeed it is their computation that makes them useful to us. Here, the snake computes where to move. While information transfer does not appear to be important for coordinated motion in flat environments, it could underpin computation for tasks such as successful navigation in challenging environments, where different parts of the body could sample many sections of the environment in parallel, and communicate information about the environment along the structure. Information transfer could be used to develop the required computational capability for tasks such as these in future work.

We intend to explore the use of information transfer in guided self-organisation in other settings where bidirectional information transfer may be required for distributed computation. We also intend to investigate the use of the other information dynamics of computation (information storage and modification) in such design, and explore the circumstances under which each should be used and indeed how they can be used together.

8.4 Summary

In this chapter, we have demonstrated the local transfer entropy to provide novel insights into biological and bio-inspired systems that the average value cannot. In Sect. 8.1 we showed that it reveals interesting detail about the dynamics in time of the heart and breath rate interaction in sleep apnea. In Sect. 8.2 we showed that localising the TE in space could reveal interregional structure in brain imaging data. Also, in Sect. 8.3 the local TE revealed the useful structure that emerged from the guided evolution of a snake-like robot.

Importantly also, we used these local insights to present and analyse new applications of the (average) transfer entropy itself. We introduced a novel approach to inferring effective network structure in the domain of computational neuroscience, which provides a unique combination of directional, non-linear, model-free, collective analysis that is applicable to small data sets. Finally, we presented the first application of the TE as an intrinsic goal in guided self-organisation, demonstrating that it can induce useful, travelling coherent information structures, akin to gliders in CAs.

References

1. T. Schreiber, Measuring information transfer. Phys. Rev. Lett. **85**(2), 461–464 (2000)
2. D.R. Rigney, A.L. Goldberger, W. Ocasio, Y. Ichimaru, G.B. Moody, R. Mark, in *Multi-channel physiological data: description and analysis*, ed. by A.S. Weigend, N.A. Gershenfeld. Time Series Prediction: forecasting the future and understanding the past (Addison-Wesley, Reading, 1993), pp. 105–129
3. M. Rubinov, S.A. Knock, C.J. Stam, S. Micheloyannis, A.W.F. Harris, L.M. Williams, M. Breakspear, Small-world properties of nonlinear brain activity in schizophrenia. Hum. Brain Mapp. **30**, 403–416 (2009)
4. R.A. Stevenson, S. Kim, T.W. James, An additive-factors design to disambiguate neuronal and areal convergence: measuring multisensory interactions between audio, visual, and haptic sensory streams using fMRI. Exp. Brain Res. **198**(2–3), 183–194 (2009)
5. K.J. Friston, Functional and effective connectivity in neuroimaging: a synthesis. Hum. Brain Mapp. **2**, 56–78 (1994)
6. C.J. Honey, R. Kotter, M. Breakspear, O. Sporns, Network structure of cerebral cortex shapes functional connectivity on multiple time scales. in *Proceedings of the National Academy of Sciences*, vol. 104, no.24, pp. 10240–10245. 2007
7. C.S. Soon, M. Brass, H.-J. Heinze, J.-D. Haynes, Unconscious determinants of free decisions in the human brain. Nat. Neurosci. **11**(5), 543–545 (2008)
8. S. Bode, J.-D. Haynes, Decoding sequential stages of task preparation in the human brain. NeuroImage **45**(2), 606–613 (2009)
9. S.L. Bressler, W. Tang, C.M. Sylvester, G.L. Shulman, M. Corbetta, Top-down control of human visual cortex by frontal and parietal cortex in anticipatory visual spatial attention. J. Neurosci. **28**(40), 10056–10061 (2008)
10. A. Tang, C. Honey, J. Hobbs, A. Sher, A. Litke, O. Sporns, J. Beggs, Information flow in local cortical networks is not democratic. BMC Neurosci. **9**(Suppl 1), O3 (2008)
11. H. Hinrichs, H.J. Heinze, M.A. Schoenfeld, Causal visual interactions as revealed by an information theoretic measure and fMRI. NeuroImage **31**(3), 1051–1060 (2006)
12. B. Chai, D. B. Walther, D. M. Beck, L. Fei-Fei, in *Exploring Functional Connectivity of the Human Brain Using Multivariate Information Analysis*, ed. by Y. Bengio, D. Schuurmans, J. Lafferty, C.K.I. Williams, A. Culottain. Advances in Neural Information Processing Systems, vol. 22, pp. 270–278 (NIPS Foundation, San Diego, 2009)
13. H. Liang, M. Ding, S.L. Bressler, Temporal dynamics of information flow in the cerebral cortex. Neurocomputing **38–40**, 1429–1435 (2001)
14. J.T. Lizier, J-D. Haynes, J. Heinzle, M. Prokopenko, Directed information structure in inter-regional cortical interactions in a visuomotor tracking task. in *Proceedings of the Eighteenth Annual Computational Neuroscience Meeting Computational Neuroscience 2009 (CNS*2009)*, BMC Neuroscience, 10(Suppl 1) (Germany, Berlin, 2009), p. P117
15. J.T. Lizier, J. Heinzle, A. Horstmann, J.-D. Haynes, M. Prokopenko, Multivariate information-theoretic measures reveal directed information structure and task relevant changes in fMRI connectivity. J. Comput. Neurosci. **30**(1), 85–107 (2011)
16. K. Young, Y. Chen, J. Kornak, G.B. Matson, N. Schuff, Summarizing complexity in high dimensions. Phys. Rev. Lett. **94**(9), 098701 (2005)
17. K. Young, N. Schuff, Measuring structural complexity in brain images. NeuroImage **39**(4), 1721–1730 (2008)
18. K.A. Norman, S.M. Polyn, G.J. Detre, J.V. Haxby, Beyond mind-reading: multi-voxel pattern analysis of fMRI data. Trends Cognitive Sci. **10**(9), 424–430 (2006)
19. A.S. Klyubin, D. Polani, C.L. Nehaniv, Representations of space and time in the maximization of information flow in the perception-action loop. Neural Comput. **19**(9), 2387–2432 (2007)
20. A. Kraskov, H. Stögbauer, P. Grassberger, Estimating mutual information. Phys. Rev. E **69**(6), 066138 (2004)

21. A. Kraskov, Synchronization and Interdependence Measures and their Applications to the Electroencephalogram of Epilepsy Patients and Clustering of Data. Ph.D. thesis, ser. Publication Series of the John von Neumann Institute for Computing, vol. 24 (John von Neumann Institute for Computing, Jülich, 2004)
22. S. Frenzel, B. Pompe, Partial mutual information for coupling analysis of multivariate time series. Phys. Rev. Lett. **99**(20), 204101 (2007)
23. G. Gomez-Herrero, W. Wu, K. Rutanen, M.C. Soriano, G. Pipa, R. Vicente, Assessing coupling dynamics from an ensemble of time series (2010), arXiv:1008.0539, http://arxiv.org/abs/1008.0539. Accessed 2010
24. M. Chávez, J. Martinerie, M. Le Van Quyen, Statistical assessment of nonlinear causality: application to epileptic EEG signals. J. Neurosci. Methods **124**(2), 113–128 (2003)
25. M. Grosse-Wentrup, in *Understanding brain connectivity patterns during motor imagery for brain-computer interfacing*, ed by D. Koller, D. Schuurmans, Y. Bengio, L. Bottou. Advances in Neural Information Processing Systems, vol. 21 (Curran Associates, New York, 2008), pp. 561–568
26. T.Q. Tung, T. Ryu, K.H. Lee, D. Lee, Inferring gene regulatory networks from microarray time series data using transfer entropy, ed by P. Kokol, V. Podgorelec, D. Mičetič-Turk, M. Zorman, M. Verlič. in *Proceedings of the Twentieth IEEE International Symposium on Computer-Based Medical Systems (CBMS '07), Maribor, Slovenia* (IEEE, Los Alamitos, USA, 2007), pp. 383–388
27. J.V. Haxby, M.I. Gobbini, M.L. Furey, A. Ishai, J.L. Schouten, P. Pietrini, Distributed and overlapping representations of faces and objects in ventral temporal cortex. Science **293**(5539), 2425–2430 (2001)
28. M. Lungarella, T. Pegors, D. Bulwinkle, O. Sporns, Methods for quantifying the informational structure of sensory and motor data. Neuroinformatics **3**(3), 243–262 (2005)
29. P.F. Verdes, Assessing causality from multivariate time series. Phys. Rev. E **72**(2), 026 222–0262229 (2005)
30. A. Horstmann, Sensorimotor integration in human eye-hand coordination: neuronal correlates and characteristics of the system, Ph.D. Dissertation (Ruhr-Universität Bochum, Bochum, 2008)
31. P. Shannon, A. Markiel, O. Ozier, N.S. Baliga, J.T. Wang, D. Ramage, N. Amin, B. Schwikowski, T. Ideker, Cytoscape: a software environment for integrated models of biomolecular interaction networks. Genome Res. **13**(11), 2498–2504 (2003)
32. P. Gong, C. van Leeuwen, Distributed dynamical computation in neural circuits with propagating coherent activity patterns. PLoS Comput. Biol. **5**(12), e1000611 (2009)
33. M. Prokopenko, V. Gerasimov, I. Tanev, Evolving spatiotemporal coordination in a modular robotic system, ed. by S. Nolfi, G. Baldassarre, R. Calabretta, J. Hallam, D. Marocco, J.-A. Meyer, D. Parisiin. *Proceedings of the Ninth International Conference on the Simulation of Adaptive Behavior (SAB'06), Rome*, ser. Lecture Notes in Artificial Intelligence, vol. 4095 (Springer, Heidelberg, 2006), pp. 548–559
34. D. Polani, O. Sporns, and M. Lungarella, How information and embodiment shape intelligent information processing, ed. by M. Lungarella, F. Iida, J. Bongard, R. Pfeifer. in *Proceedings of the 50th Anniversary Summit of Artificial Intelligence, New York*, ser. Lecture Notes in Computer Science, vol. 4850 (Springer, Berlin, 2007), pp. 99–111
35. A. S. Klyubin, D. Polani, C. L. Nehaniv, All else being equal be empowered, ed. by M.S. Capcarrere, A.A. Freitas, P.J. Bentley, C.G. Johnson, J. Timmis. in *Proceedings of the 8th European Conference on Artificial Life (ECAL), Kent, UK*, ser. Lecture Notes in Computer Science, vol. 3630 (Springer, Heidelberg, 2005), pp. 744–753
36. O. Sporns, M. Lungarella, Evolving coordinated behavior by maximizing information structure, ed. by L.M. Rocha, L. S. Yaeger, M. A. Bedau, D. Floreano, R. L. Goldstone, A. Vespignani. in *Proceedings of the Tenth International Conference on Simulation and Synthesis of Living Systems (ALifeX), Bloomington, Indiana, USA* (MIT Press, Cambridge, 2006), pp. 323–329
37. N. Ay, N. Bertschinger, R. Der, F. Güttler, E. Olbrich, Predictive information and explorative behavior of autonomous robots. Eur. Phys. J. B **63**(3), 329–339 (2008)

38. J.T. Lizier, M. Prokopenko, I. Tanev, A.Y. Zomaya, Emergence of glider-like structures in a modular robotic system, ed. by S. Bullock, J. Noble, R. Watson, M.A. Bedau. in *Proceedings of the Eleventh International Conference on the Simulation and Synthesis of Living Systems (ALife XI), Winchester, UK*, (MIT Press, Cambridge, 2008), pp. 366–373
39. M. Prokopenko, V. Gerasimov, I. Tanev, Measuring spatiotemporal coordination in a modular robotic system ed. by L.M. Rocha, L.S. Yaeger, M.A. Bedau, D. Floreano, R.L. Goldstone, A. Vespignani. in *Proceedings of the 10th International Conference on the Simulation and Synthesis of Living Systems (ALifeX), Bloomington, Indiana, USA* (MIT Press, Cambridge, 2006), pp. 185–191
40. I. Tanev, T. Ray, A. Buller, Automated evolutionary design, robustness, and adaptation of sidewinding locomotion of a simulated snake-like robot. IEEE Trans. Robot. **21**(4), 632–645 (2005)
41. J.T. Lizier, M. Prokopenko, A.Y. Zomaya, Local information transfer as a spatiotemporal filter for complex systems. Phys. Rev. E **77**(2), 026110 (2008)
42. J.A. Brown, J.A. Tuszynski, A review of the ferroelectric model of microtubules. Ferroelectrics **220**, 141–156 (1999)
43. I. Couzin, R. James, D. Croft, J. Krause, in *Social organization and information transfer in schooling fishes*, ed by B.C.K. Laland, J. Krause. Fish Cognition and Behavior, ser. Fish and Aquatic Resources (Blackwell Publishing, Oxford, 2006), pp. 166–185
44. M. Mitchell, J.P. Crutchfield, P.T. Hraber, Evolving cellular automata to perform computations: mechanisms and impediments. Physica D **75**, 361–391 (1994)
45. M. Mitchell, J.P. Crutchfield, R. Das, Evolving cellular automata with genetic algorithms: a review of recent work ed. by E.D. Goodman, W. Punch, V. Uskov. in *Proceedings of the First International Conference on Evolutionary Computation and Its Applications, Moscow* (Russian Academy of Sciences, Russia, 1996)

Chapter 9
Conclusion

This thesis has presented a framework for the information dynamics of distributed computation in complex systems. We summarise the main contributions of this work in Sect. 9.1, and suggest directions for future exploration in Sect. 9.2.

9.1 Summary of Main Contributions

The results presented in this thesis have upheld the hypothesis in Sect. 1.1 that *with the ability to describe and locally quantify distributed computation in terms of information storage, transfer and modification, we will be better able to understand distributed computation in nature and its sources of complexity*. In this section, we describe our contribution to the fundamental understanding of distributed computation in complex systems.

9.1.1 Framework for the Information Dynamics of Distributed Computation

The primary contribution of this thesis is the first *complete framework to quantify the information dynamics of distributed computation*. That is, the framework quantifies computation in terms of the component operations on information: storage, transfer and modification. The framework has a particular focus on the dynamics of these operations on a local scale in space and time within a system.

There are three key properties of the framework which underpin its novelty:

1. With an **information-theoretic basis**, the framework captures non-linear effects and is applicable to any type of dynamic process (i.e. including both discrete and continuous valued states);

J. T. Lizier, *The Local Information Dynamics of Distributed Computation in Complex Systems*, Springer Theses, DOI: 10.1007/978-3-642-32952-4_9,
© Springer-Verlag Berlin Heidelberg 2013

2. The approach is directly relevant to the language in which distributed computation in complex systems is normally described, via the concepts of **memory, communications and processing**, since these directly map to the operations of information storage, transfer and modification;
3. The framework focuses on the **local dynamics** of these operations on information. While averaged or system-wide measures have their place in providing summarised results, the local focus in space and time is vital for understanding the nature of each measure and providing insights about system behaviour that averaged measures cannot.

In the next three Sects. (9.1.2–9.1.4), we will describe the specific contributions made regarding each individual operation on information. Many of the measures presented in this framework are original contributions; all of the measures discussed are examined on a local scale here for the first time.

Incorporating the measures together in a single framework allowed us to provide insights that would not be possible with separate investigation. For example, we demonstrated how the component operations interrelate in the computation of the next state of a given variable. We also found that establishing the context of the past history of the destination was at the heart of the perspective of distributed computation, and was critical for accurate quantification of each operation on information. Also, the use of a single framework allowed us to provide broader insights into the fundamental nature of distributed computation, e.g. regarding cellular automata as described in Sect. 9.1.5.

9.1.2 Measuring Information Storage

We described how information storage is quantified in terms of either total storage (via the existing measure *excess entropy*) or the amount of storage currently in use (via the new measure *active information storage*). We presented the first *localisation* of both measures, allowing us to contrast the insights they provide on distributed computation. We also made clear the manner in which information storage in a distributed computation can be implemented using an agent's environment as the storage medium.

9.1.3 Measuring Information Transfer

We described how information transfer is quantified using the existing measure *transfer entropy*. We introduced a number of variants to this measure in order to capture subtly different concepts; most notably, we introduced the *complete transfer entropy* to measure information transfer taking into account how the given source interacts with other causal sources in acting on the destination. We described how to measure the transfer entropy on a *local scale* in time and space, allowing us to provide insights on how its parameters should be set, describe the relationship between its variants, and demonstrate its alignment with the popularly understood notion of information transfer (see Sect. 9.1.5). We also showed how properly establishing the past history of the destination was critical to separating information storage and transfer.

Additionally, we described the differentiation between the concepts of information transfer and causal information flow. The distinctions revealed here are particularly pertinent because of the large degree of confusion surrounding these concepts in the literature. This result was only possible using our local perspective, which including localising the existing *information flow* measure.

9.1.4 Measuring Information Modification

We described how information storage and transfer are combined in the operation of information modification, and introduced the measure *separable information* to quantitatively identify non-trivial information modification events on a local scale within a system.

We also outlined how to quantify information destruction within a distributed system, introducing a measure for *information destruction*. Using localisations of both these measures, we described the distinction between the concepts of information modification and destruction.

9.1.5 Quantitative Understanding of Information Dynamics in CAs

Our framework provided the first direct quantitative evidence for several important long-held conjectures regarding the facilitation of computation in cellular automata (CAs) via emergent structures. That is, we showed that blinkers implement information storage, moving particles (gliders and domain walls) are dominant information transfer agents, and particle collisions are information modification events. This demonstrated that our quantitative framework aligned with the popularly-understood concepts of memory, communication and processing.

CAs are a critical proving ground for any theory regarding the nature of distributed computation [1, 2], and these results suggest significant implications for our fundamental understanding of distributed computation and the dynamics of complex systems. The importance of our application to CAs is underlined because many other natural and artificial systems have been observed to process information using similar emergent coherent structures [3, 4]. We also emphasise that these insights were only possible using the *local* perspective introduced for these concepts here.

The application of the framework to CAs aligned well with other methods of spatiotemporal *filtering* for complex structure (e.g. [5–9]). Obviously, filtering of coherent structure is not a new concept. However, our work is distinct in that it provides *several different profiles of the system corresponding to each type of computational structure* (and indeed one view for each information transfer channel or direction). This approach allows more refined filtering, and is unique in providing quantitative evidence regarding the computational role of these emergent structures. Furthermore, this approach is distinct in that it: provides continuous rather than discrete values (like [7] and [9]); does not follow an arbitrary spatial preference (unlike [6] and [9]) but rather the flow of time only; is automatically obtained (unlike the approaches manually crafted for specific CA rules in [5, 10]) and like [7] does not require a new filter for every CA (though the probability distribution functions must be recalculated individually). Finally, by focussing on these operations on information it highlights subtly different parts of emergent structure to other filters (generally highlighting less of the structure).

9.1.6 Measuring Computational Properties in Phase Transitions in Networks

We presented the first analysis of information dynamics in order-chaos phase transitions in networks, finding that information storage and transfer were *maximised near the critical phase* of two different network types. While the finite-sized systems exhibited approximate phase transitions, we described reasons why this interesting result might be expected to be generalised in similar[1] order-chaos phase transitions.

The results were of particular interest because of the network models chosen for study: random Boolean networks (RBNs, a model of gene regulatory networks) and a model of cascading failures. In particular, the use of RBNs was important as they had been a focus for conjecture that computational properties were maximised near the critical phase (in alignment with the *edge of chaos* hypothesis). More generally, we revealed several interesting ways in which underlying network topology drives information dynamics. Indeed, several leading authors in network science suggest that dynamics are the next frontier in this domain [11–13], so the promising results from information dynamics here are notable in suggesting it as a generally-applicable candidate for further investigation.

9.1.7 Methodology for Studying Coherent Information Structure

We also demonstrated that the maximisation of information storage and transfer near order-chaos phase transitions is not a universal result that can be expected from any type of system exhibiting ordered and chaotic variants (in particular CAs).

[1] Similar phase transitions being those with transitions controlled by a single order-chaos parameter, whether those transitions cause a discontinuous or smooth change in properties.

Instead, we observed that coherent information structure is a defining feature of complex computation and presented a *methodology for studying coherent information structure*. Importantly, our approach identifies both clear and "hidden" coherent structure in complex computation, most notably reconciling conflicting interpretations of the complexity in CA rule 22.

9.1.8 Demonstrated Application Areas for Information Dynamics

Finally, we demonstrated the utility of the framework for information dynamics in two key application areas. Together, these showed its flexibility to different data types and ability to provide useful insights to practical problems.

We presented a *method for inferring directed interregional information structure* in multivariate data sets, e.g. time-series brain imaging data. The method is unique in combining the features of directional, non-linear, model-free analysis, on a regional level, capturing the results of interaction of multiple sources, and being robust to relatively small data sets. We demonstrated the efficacy of the method by applying it to an fMRI data set, revealing a tiered information structure that correlates well with the cognitive task the subjects were performing.

We also reported the first use of the transfer entropy as a fitness function for *guiding self-organisation*. This example demonstrated that the approach can induce the emergence of useful coherent information structure in a system, which could only be revealed by examining local information dynamics.

9.2 Directions for Future Work

Certainly there are ways in which **the experiments reported here can be directly expanded** for deeper analysis. For example, we described in Sect. 6.3 several ways in which the analysis of the phase transition in RBNs could be expanded, including investigating the effects of noise. There are also several directions outlined in Sect. 7.4 in which our analysis of coherent information structure should be pursued further, in particular in examining other measures of structure in the information state-space and the relationship of coherent information structure to overall complexity.

In addition to revisiting these experiments though, new work is required to build on the achievements of this thesis in both theoretical and practical directions.

Arguably the most important direction for *theoretical work* is to investigate the **relationship between the topology of networks and their information dynamics**. As discussed in Chap. 6, most leading authors in network science suggest that [11–13] the next great leaps in that field will be produced from understanding *time-series dynamics* and how they are coupled with network topology. We described the manner in which the research landscape suggests that the dynamics of distributed computation has the potential to be the widely-anticipated framework of choice for the study of time-series dynamics in networks. Two key reasons for this are: that the approach is generic and can be applied to any type of time-series dynamics; and that the language of computation pervades description of the time-series dynamics of networks. We have demonstrated important preliminary results in applying our framework to analyse computation in networks in Chap. 6. More work is required here though, in particular a thorough investigation of how information dynamics are imparted from underlying network topology. Such investigations will establish for example whether special topologies such as small-world and scale-free networks are distinguished from others by their computational properties, and how local topological structures [14] relate to local computational capabilities.

Further theoretical work is also required to establish **how the framework for information dynamics relates to other approaches and fields**, and to clarify the measurement of several important concepts in distributed computation. A primary example here is to establish the relationship with ϵ-machines and statistical complexity from *computational mechanics* [7, 15–19]. Certainly the relationship between excess entropy and statistical complexity is well-established [18, 20–22] (indeed the excess entropy originated in computational mechanics). The perspective of *distributed* computation here would be novel in considering how information storage and transfer together related to the overall statistical complexity. This would involve focussing on the light-cone formulation of computational mechanics, which considers how the next state of an agent (and its causal descendants) depends on the causal contributors to that agent [17]. Another interesting direction for exploration would be to examine whether the framework for distributed computation can be usefully altered to apply to the underlying internal causal states of the variables in a distribution computation. Similarly, just as computational mechanics is exploring measuring information storage in quantum computation (e.g. the quantum excess entropy [23]), our framework should be extended for application to *distributed quantum computation*. Also, we note work considering quantifying *interaction structures* [24]: i.e. investigating kth order statistical dependencies between variables that cannot be reduced to dependencies between $k - 1$ of them. New work is required to quantify similar interaction structures in the context of distributed computation (i.e. examining how many source information sources are irreducibly interacting to produce an outcome). This work should also establish how this is related to the distributed operations on information (especially information modification), and whether the concept can be quantified on a local scale in space and time.

We have demonstrated a number of promising practical results in the applications of the framework to date demonstrated in this thesis (e.g. in Chap. 8). That being said, there is much more scope for quantifying computation and producing both interesting and useful insights in **applying the framework in practical settings**. Such applications will not only provide useful insights in the domain under consideration, but also build momentum for further use of the approach.

A key application area will be **computational neuroscience**. In this domain there is an abundance of time-series imaging data, and powerful capability for computational analysis, yet the road forward to specifically understand distributed computation in the brain is unclear. As explored in Sect. 8.2, information dynamics offers potential for ground-breaking insights in providing analysis of space-time information patterns, and revealing how the brain is computing. We have demonstrated utility of the approach in this context by revealing directed information structure supporting a cognitive task in Sect. 8.2; future work will include applying our method to other cognitive tasks. We will also examine other data types, particularly those with shorter time scales which allow more direct conclusions about neuronal interactions. Furthermore, we will seek to expand the application of information dynamics here, in particular in examining information storage and modification in addition to transfer. Similarly, we will examine the information dynamics on a *local scale in time* as well as space in brain-imaging data. This could lead to the identification of travelling coherent information structures in the cortex (as described in [25]). The local perspective will also address questions such as "how much information is transferred from region A to B *at time* t?", specifically revealing the information dynamics associated with particular cognitive tasks. We will explore whether this direct approach to revealing space-time information interactions can improve on inferences of shared information such as those in [26]. Additionally, we will investigate the use of the framework to infer effective networks [27] on the level of individual variables (e.g. voxels) rather than regions. Building on the use of the transfer entropy alone (e.g. [28, 29]), this could be performed using the inference method described in Appendix E to determine the sources contributing to a node's computation of its next state.

Another important application area will be in **guiding self-organisation**. As described in Sects. 2.5 and 8.3, we that a promising approach to this type of system design is the use of measures of the information dynamics of distributed computation. This is primarily because any task we wish the system to solve involves a distributed computation, so focussing our guidance on providing the fundamental building blocks of the computation is a direct way to allow that computation to emerge. We reported preliminary results indicating that evolving to maximise information transfer on local links can lead to the emergence of useful coherent information structure on a global level. Future work will include examining the use of information dynamics to guide other types of self-organised systems, e.g. collective motion or flocking. An interesting domain will be examining how information dynamics can guide network topology, for example whether our understanding of the information dynamics of cascading failures in Chap. 6 can be applied to design power grids to avoid these events. From a theoretical perspective, this scope of our work in guided self-organisation needs to be significantly expanded to consider information storage and modification also, and to establish what types of properties can usefully produced by processes of evolution or adaptation to maximise each of them. More importantly, the approach needs to investigate how the information dynamics can be used *together* to guide the emergence of universal computation.[2] Intricate tasks will require such arbitrarily-complex computation (facilitated by bidirectional

[2] Indeed, whether these measures can be used to determine the capability of a distributed system for universal computation, or capability of other levels of computational complexity [16], needs to be established.

information transfer, storage structures and modification events) as distinct from computation that exhibits only one type of operation.

References

1. M. Mitchell, in *Computation in cellular automata selected review*, ed. by T. Gramss, S. Bornholdt, M. Gross, M. Mitchell, T. Pellizzari, non-standard computation, (Verlagsgesellschaft, Weinheim, 1998), pp. 95–140
2. J. Von Neumann, *Theory of self-reproducing automata* ed. by A.W. Burks (University of Illinois Press, Urbana, 1966)
3. D. Peak, J.D. West, S.M. Messinger, K.A. Mott, Evidence for complex, collective dynamics and emergent, distributed computation in plants. Proc. Nat. Acad. Sci. USA. **101**(4), 918–922 (2004)
4. M. Mitchell, J.P. Crutchfield, P.T. Hraber, Evolving cellular automata to perform computations: mechanisms and impediments. Physica D **75**, 361–391 (1994)
5. P. Grassberger, New mechanism for deterministic diffusion. Phys. Rev. A **28**(6), 3666 (1983)
6. J.E. Hanson, J.P. Crutchfield, The attractor-basin portait of a cellular automaton. J. Stat. Phys. **66**, 1415–1462 (1992)
7. C.R. Shalizi, R. Haslinger, J.-B. Rouquier, K.L. Klinkner, C. Moore, Automatic filters for the detection of coherent structure in spatiotemporal systems. Phys. Rev. E **73**(3), 036104 (2006)
8. A. Wuensche, Classifying cellular automata automatically: finding gliders, filtering, and relating space-time patterns, attractor basins, and the Z parameter. Complex **4**(3), 47–66 (1999)
9. T. Helvik, K. Lindgren, M.G. Nordahl, Local information in one-dimensional cellular automata, in *Proceedings of the International Conference on Cellular Automata for Research and Industry*, ed. by P.M. Sloot, B. Chopard, A.G. Hoekstra. Amsterdam, ser. Lecture Notes in Computer Science, vol. 3305 (Springer, Berlin, 2004), pp. 121–130
10. P. Grassberger, Information content and predictability of lumped and distributed dynamical systems. Phys. Scr. **40**(3), 346 (1989)
11. D.J. Watts, *Six Degrees: The Science of a Connected Age* (Norton, New York, 2003)
12. A.-L. Barabási, Scale-free networks: a decade and beyond. Science **325**(5939) 412–413 (2009)
13. M. Mitchell, Complex systems: Network thinking. Artif. Intell. **170**(18), 1194–1212 (2006)
14. M. Piraveenan, M. Prokopenko, A.Y. Zomaya, Local assortativeness in scale-free networks. Europhys. Lett. **84**(2), 28002 (2008)
15. J.P. Crutchfield, K. Young, Inferring statistical complexity. Phys. Rev. Lett. **63**(2), 105 (1989)
16. J.P. Crutchfield, The calculi of emergence: computation, dynamics and induction. Physica D **75**(1–3), 11–54 (1994)
17. C.R. Shalizi, *Causal architecture, complexity and self-organization in time series and cellular automata*, Ph.D. Dissertation, University of Wisconsin-Madison, 2001
18. C.R. Shalizi, J.P. Crutchfield, Computational mechanics: pattern and prediction, structure and simplicity. J. Stat. Phys. **104**, 817–879 (2001)
19. C.R. Shalizi, K.L. Shalizi, R. Haslinger, Quantifying self-organization with optimal predictors. Phys. Rev. Lett. **93**(11), 118701 (2004)
20. J.P. Crutchfield, C.J. Ellison, J.R. Mahoney, Time's barbed arrow: irreversibility, crypticity, and stored information. Phys. Rev. Lett. **103**(9), 094101 (2009)
21. C. Ellison, J. Mahoney, J. Crutchfield, Prediction, retrodiction, and the amount of information stored in the present. J. Stat. Phys. **136**(6), 1005–1034 (2009)
22. K. Wiesner, M. Gu, E. Rieper, V. Vedral, Information erasure lurking behind measures of complexity, 2009, arXiv:0905.2918v1. url: http://arxiv.org/abs/0905.2918
23. J.P. Crutchfield, K. Wiesner, Intrinsic quantum computation. Phys. Lett. A **372**(4), 375–380 (2008)

24. T. Kahle, E. Olbrich, J. Jost, N. Ay, Complexity measures from interaction structures. Phys. Rev. E **79**(2), 026201 (2009)
25. P. Gong, C. van Leeuwen, Distributed dynamical computation in neural circuits with propagating coherent activity patterns. PLoS Comput. Biol. **5**(12), e1000611 (2009)
26. C.S. Soon, M. Brass, H.-J. Heinze, J.-D. Haynes, Unconscious determinants of free decisions in the human brain. Nat. Neurosci. **11**(5), 543–545 (2008)
27. K.J. Friston, Functional and effective connectivity in neuroimaging: a synthesis. Hum. Brain Mapp. **2**, 56–78 (1994)
28. C.J. Honey, R. Kotter, M. Breakspear, and O. Sporns, Network structure of cerebral cortex shapes functional connectivity on multiple time scales, in Proc. Nat. Acad. Sci. **104**(24), 10240–10245 (2007)
29. J.T. Lizier, M. Piraveenan, D. Pradhana, M. Prokopenko, L. S. Yaeger, Functional and structural topologies in evolved neural networks, in *Proceedings of the European Conference on Artificial Life (ECAL)*, ed. by G. Kampis, I. Karsai, E. Szathmáry. Budapest, Hungary, ser. Lecture Notes in Computer Science, vol. 5777 (Springer, Berlin, 2011), pp. 140–147

Appendix A
Consideration of Alternative Method of Localisation

An alternative method of localising mutual information-based measures was proposed in [1]. The authors consider *partial* localisations, computing how much information $I(y_n; X)$ a specific value y_n gives about what value X *might* take. It is required that a partial localisation $I(y_n; X)$ averages over y_n to the average mutual information $I(Y; X)$:

$$I(Y; X) = \sum_{y_n} p(y_n) I(y_n; X). \tag{A.1}$$

As well as the conventional expression that satisfies this requirement:

$$I_1(y_n; X) = \sum_{x_n} p(x_n \mid y_n) \log_2 \frac{p(x_n \mid y_n)}{p(x_n)}, \tag{A.2}$$

the authors present an alternative partial local mutual information as the reduction in uncertainty of X on knowing y_n:

$$I_2(y_n; X) = H_X - H_{X|y_n}, \tag{A.3}$$

giving:

$$I_2(y_n; X) = -\sum_{x_n} p(x_n) \log_2 p(x_n) + \sum_{x_n} p(x_n \mid y_n) \log_2 p(x_n \mid y_n). \tag{A.4}$$

While both I_1 and I_2 satisfy the constraint Eq. (A.1), they do give different values for $I(y_n; X)$. Importantly, I_1 is *non-negative*, but I_2 is unique in satisfying the key

J. T. Lizier, *The Local Information Dynamics of Distributed Computation in Complex Systems*, Springer Theses, DOI: 10.1007/978-3-642-32952-4,
© Springer-Verlag Berlin Heidelberg 2013

property of *additivity* of information from multiple sources[1]:

$$I(\{y_n, z_n\}; X) = I(y_n; X) + I(z_n; X \mid y_n). \tag{A.5}$$

In this paper we consider *full* localisations, computing how much information $i(y_n; x_n)$ a value y_n gives about the specific value x_n that X *actually* takes at time step n. Similar to requirement Eq. (A.1), the full localisations $i(y_n; x_n)$ are required to satisfy:

$$I(Y; X) = \sum_{y_n} p(y_n) \sum_{x_n} p(x_n \mid y_n) i(y_n; x_n). \tag{A.6}$$

The approach to these local values used in the main body of our text (see Eq. (2.25) in Sect. 2.2.2):

$$i_1(y_n; x_n) = \log_2 \frac{p(x_n \mid y_n)}{p(x_n)}, \tag{A.7}$$

is analogous to $I_1(y_n; X)$ because it also satisfies:

$$I(y_n; X) = \sum_{x_n} p(x_n \mid y_n) i(y_n; x_n), \tag{A.8}$$

for $I_1(y_n; X)$. Interestingly, for $i_1(y_n; x_n)$ we also have:

$$i_1(y_n; x_n) = h(x_n) - h(x_n \mid y_n), \tag{A.9}$$

in analogy to $I_2(y_n; X)$ in Eq. (A.3), which leads i_1 to satisfy the crucial property of *additivity* [1]:

$$i(\{y_n, z_n\}; x_n) = i(y_n; x_n) + i(z_n; x_n \mid y_n), \tag{A.10}$$

unlike $I_1(y_n; X)$ (with Eq. (A.5)).

It is worth considering whether the approach of [1] in proposing $I_2(y_n; X)$ may be extended to propose a valid $i_2(y_n; x_n)$ which satisfies Eq. (A.6) by satisfying Eq. (A.8) for $I_2(y_n; X)$. Certainly an extension of Eq. (A.4) provides:

$$i_2(y_n; x_n) = -\frac{p(x_n)p(y_n)}{p(x_n, y_n)} \log_2 p(x_n) + \log_2 p(x_n \mid y_n), \tag{A.11}$$

for this purpose. However, this expression does not satisfy the additivity property of Eq. (A.10).

Importantly also, expressions for $i(y_n; x_n)$ have an additional requirement for correctness: they **must** be *symmetric* in x_n and y_n in analogy to the averaged value

[1] The property of additivity is also referred to as *recursion* in [2], and was one of Shannon's original requirements for the *averaged* measure H [2, 3].

$I(X; Y)$ because the information contained in y_n about the specific value x_n is the same as the information contained in x_n about the specific value of y_n. This is not applicable to partial localisations $I(y_n; X)$ because they are asymmetrically defined in considering the known value of one variable and the unknown value of the other.

The extension of $I_2(y_n; X)$ to $i_2(y_n; x_n)$ fails this symmetry requirement in general (easily verified with sample values, e.g. $p(x_n) = 0.1$, $p(y_n) = 0.18$, $p(x_n \mid y_n) = 0.5$, $p(y_n \mid x_n) = 0.9$, $p(x_n, y_n) = 0.09$), and so is not a correct form to locally quantify the mutual information.

As such, we are left with $i_1(y_n; x_n)$ for *full* localisations $i(y_n; x_n)$ since *it satisfies both additivity and symmetry*.

When selecting a measure for *partial* localisations, one should carefully consider which properties are required. Selecting $I_2(y_n; X)$ preserves additivity, while $I_1(y_n; X)$ preserves positivity and averaging over the correct full localisation $i_1(y_n; x_n)$.

References

1. M.R. DeWeese, M. Meister, How to measure the information gained from one symbol. Netw. Comput. Neural Syst. **10**, 325–340 (1999)
2. M. Prokopenko, F. Boschietti, A.J. Ryan, An information-theoretic primer on complexity, self-organization, and emergence. Complexity **15**(1), 11–28 (2009)
3. C.E. Shannon, A mathematical theory of communication. Bell Syst. Tech. J. 27, 379–423 and 623–656 (1948)

Appendix B
Entropy Rate Convergence and Divergent Excess Entropy

Grassberger studied temporal entropy rate estimates for several ECAs in [1, 2] in order to gain insights into their excess entropies. These studies estimated temporal entropy rates $H_{\mu,N}(k)$ for spatial blocks of size N as N is increased. Estimated values of $H_{\mu,N}(k)$ (for $N = 1$ and in the limit as $N \rightarrow \infty$) were catalogued for most ECAs in a table of statistical properties in [3]. Using these estimates, the studies focused on inferring the collective excess entropies rather than the single-agent ($N = 1$) excess entropies. For several rules (including rule 22, studied with Monte Carlo estimates), the temporal entropy rate estimates $H_{\mu,N}(k)$ (for $N > 1$) were concluded to follow a power law decay to their asymptote $H_{\mu,N}$:

$$H_{\mu,N}(k) = H_{\mu,N} + C/k^{\alpha}, \tag{B.1}$$

with exponent $\alpha \leq 1$ (C is a constant). This is significant because with $\alpha \leq 1$ the collective excess entropy (known as *effective measure complexity* in [1]) is divergent, implying a highly complex process. This case has been described as "a phenomenon which can occur in more complex environments", as with strong long-range correlations a semi-infinite sequence "could store an infinite amount of information about its continuation" [4] (as per the predictive information form of the excess entropy Eq. (2.19)). Rule 22 was inferred to have $H_{\mu,N} = 0$ and infinite excess entropy, which can be interpreted as a process requiring an infinite amount of memory to maintain an aperiodicity [5]. Indeed, Grassberger states that "very long-range correlations are necessary to prevent the distribution from collapsing to something periodic." [2]. Alternative methods for computing two-dimensional excess entropies, which would be applicable for computing the collective excess entropy in CAs, were presented by Feldman and Crutchfield in [6].

In attempting to quantify *local* information dynamics of distributed computation here, our focus is on information storage for *single agents or cells* rather than the joint information storage across the collective. Were such power-law trends to exist for the single-agent case, they may be more significant than for the collective case. Infinite collective excess entropy may only imply that the collective is at least trivially utilising all of its available memory (note that even the chaotic rule 30 exhibits

J. T. Lizier, *The Local Information Dynamics of Distributed Computation in Complex Systems*, Springer Theses, DOI: 10.1007/978-3-642-32952-4,

© Springer-Verlag Berlin Heidelberg 2013

divergence). For example, a CA (with infinite width) that simply copied cell values to the right would have infinite collective excess entropy when started from random initial states, yet this is clearly a trivial use of this storage. On the other hand, divergent single-agent excess entropy would imply that all agents are individually strongly utilising the resources of the collective in a highly complex process. One could go on to study the entropy rate convergence for single agents ($N = 1$),[2] however any findings would be subject to the problems with overall or averaged measures described earlier. We hypothesise that local measures in *time* as well as space will provide more detailed insights into the computation taking place in CAs.

References

1. P. Grassberger, Toward a quantitative theory of self-generated complexity. Int. J. Theor. Phys. **25**(9), 907–938 (1986)
2. P. Grassberger, Long-range effects in an elementary cellular automaton. J. Stat. Phys. **45**(1–2), 27–39 (1986)
3. P. Grassberger, in Table 6: Statistical Properties, ed by S. Wolfram. Theory and Applications of Cellular Automata (World Scientific Publishing Co. Ltd, Singapore, 1986).
4. K. Lindgren, M.G. Nordahl, Complexity measures and cellular automata. Complex Syst. **2**(4), 409–440 (1988)
5. J.P. Crutchfield, D.P. Feldman, Regularities unseen, randomness observed: levels of entropy convergence. Chaos **13**(1), 25–54 (2003)
6. D.P. Feldman, J.P. Crutchfield, Structural information in two-dimensional patterns: entropy convergence and excess entropy. Phys. Rev. E **67**(5), 051104 (2003)
7. A. Clauset, C.R. Shalizi, M.E.J. Newman, Power-law distributions in empirical data. SIAM Rev. **51**(4), 661–703 (2009)

[2] In doing so, one should use proper quantitative techniques for establishing the presence or otherwise of power-laws, similar to [7] but tailored to study measurements of general functions rather than distributions only.

Appendix C
Relation of Transfer Entropy to Massey's Directed Information

In this appendix, we describe the relationship between the transfer entropy and Massey's directed information [1]. The directed information is the *sum* of information gained about each time step of the destination x_{n+1} from the *concurrent and all past states* of the source $y_{n+1}^{(n)}$, that was not contained in the past states of the destination $x_n^{(n-1)}$ (capped over an N time step sequence):

$$I_{Y \to X}(N) = \sum_{n=1}^{N} I_{Y^{(n)\prime}; X' | X^{(n-1)}}, \tag{C.1}$$

While the transfer entropy and the directed information are obviously quite similar, there are several interesting differences between them. In particular, the directed information looks at the *sum* of information gained over the time series, rather than the average at each time step. Also, the directed information looks at all available information from the past of the source (akin to setting l to ∞ in the transfer entropy, see Eq. (4.1)) rather than only 1 state or limiting to consider only directly causal sources (as we recommend for the transfer entropy in Sect. 4.4.2). We see that the directed information approaches the question of how *two series as a whole* relate to each other rather than the relationship between corresponding *individual values*. Where the transfer entropy is differentiated as directed and dynamic against the mutual information between individual source and destination values, the directed information is differentiated as directed and dynamic against the mutual information between the two series as a whole (as two joint variables).

Furthermore, we note that the transfer entropy uses a one step difference in time from the source to the destination, while the directed information considers concurrent values of the source and destination. This is the key difference, because the time difference means the transfer entropy has physical meaning in examining how much of the source process was involved in determining the outcome of the destination: with concurrent source and destination values, the measure will detect correlations only. Where one is considering an isolated channel (in traditional information theory applications such as Massey presents), it is generally understood that concurrent

J. T. Lizier, *The Local Information Dynamics of Distributed Computation in Complex Systems*, Springer Theses, DOI: 10.1007/978-3-642-32952-4,
© Springer-Verlag Berlin Heidelberg 2013

values mean the source is a causal information contributor to the destination. However, when one moves to large multivariate systems (e.g. fMRI measurements of the cortex) with cross-causality or feedback between the variables, it is not appropriate to represent concurrent values as being causal in both directions. One could simply insert a time difference into the directed information, but it does not appear to be intended to operate along these lines.

Indeed, one could adjust the directed information to account for all of the differences outlined here, but together they imply that the measures are firmly intended to capture fundamentally different concepts. The directed information considers the extent to which one series is determined from another; the transfer entropy considers the extent to which values of a source are directly involved in a computation at the destination and in this sense has important physical meaning.

Reference

1. J.L. Massey, Causality, feedback and directed information. in *Proceedings of the International Symposium on Information Theory and its Applications* (Waikiki, Hawaii, USA, 1990).

Appendix D
Back-Door Adjustment

Certain cases exist where one can construct interventional probability distributions $p(y \mid \hat{x})$ from observational probabilities only [1]. For example, the "back-door adjustment" (Sect. 3.3.1 of [2])[3] is an option where a set of nodes U satisfies the "back-door criteria" relative to (X, Y), i.e. that:

1. no node in U is a causal descendant of X, and
2. U blocks every "back-door path" between X and Y. A back-door path between X and Y is a path of causal links connecting these nodes, where the individual links in the path may point in either direction, so long as the path includes a causal link directly into X. (See footnote 18 on p. 102 in Chap. 4 for the definition of blocking a path).

In that case, the interventional conditional probability $p(y \mid \hat{x})$ is given by:

$$p(y \mid \hat{x}) = \sum_u p(y \mid x, u) p(u). \tag{D.1}$$

The back-door adjustment could be applied to $p(a \mid \hat{s})$ in ECAs in Fig. 4.5 with the set of nodes satisfying the back-door criteria marked there as u; for $p(b \mid \hat{a}, \hat{s})$ the set $u_2 = \{u, x_{i-2,n-1}\}$ would be used. In general, note that the back-door adjustment can only be applied if all relevant combinations are observed (i.e. for (y, x, u) where $p(y, x, u)$ is strictly positive [1]). Importantly, in determining $p(a \mid \hat{s})$ for example if one does not observe all combinations $\{a, s\}$ then one cannot observe all combinations $\{a, s, u\}$ either.

---

[3] The back-door adjustment is a sub-case of the "adjustment for direct causes" [1] which is numerically simpler when the set of back-door nodes U is known.

J. T. Lizier, *The Local Information Dynamics of Distributed Computation in Complex Systems*, Springer Theses, DOI: 10.1007/978-3-642-32952-4,
© Springer-Verlag Berlin Heidelberg 2013

References

1. N. Ay, D. Polani, Information flows in causal networks. Adv. Complex Syst. **11**(1), 17–41 (2008)
2. J. Pearl, *Causality: Models, Reasoning, and Inference* (Cambridge University Press, Cambridge, 2000)

Appendix E
Complete Transfer Entropy for Causal Structure Inference

As demonstrated in Sect. 4.4.3, under certain conditions the complete transfer entropy converges with the information flow. Where one cannot intervene in the system, and does not have the required observations to use a method such as the back-door adjustment (see Appendix D), the local complete transfer entropy could provide a useful inference for the local information flow profile. Within one's control is to set the history length k to include only the past states of the destination that are causal information contributors to its next state. The history length parameter k therefore has an important role in moving the (complete) transfer entropy between measuring information transfer (at large k) and approximating causal effect (at minimal k). Outside of one's control is whether the other conditions described in Sect. 4.4.3 are met; errors begin to be introduced where they are not. We note that there is a wide class of systems where the source a is causally independent of the other causal contributors to the destination s (i.e. $p(a \mid \hat{s}) \equiv p(a)$), and though error-prone a subsequent assumption of conditional independence (i.e. $p(a \mid s) = p(a)$) is a maximum entropy assumption.

Importantly, the complete transfer entropy must condition on the correct neighbourhood of causal sources. This knowledge is missing in the important application where one is *inferring* causal structure in a multivariate time series. It is possible that the transfer entropy itself could be used to iteratively *build* an inference of the causal contributors for a given destination by *incrementally* conditioning on previously inferred sources (reminiscent of Eq. (4.47)). This would be done by incrementally identifying the next source which provides the most statistically significant transfer entropy conditioned on the previously identified sources, until all (deterministic) information in the destination is accounted for. Such a method combines the multivariate source selection of [1] with the complete transfer entropy and the statistical significance tests of [2]. Testing this method is left for future work.

Finally, we note that while the complete transfer entropy can at least function in the absence of observations spanning all possible combinations of the variables (unlike information flow), if crucial combinations are not observed it can give quite incorrect inferences here. For example, consider the classical causal example of a short circuit which causes a fire in the presence of certain conditions (e.g. with inflammable

J. T. Lizier, *The Local Information Dynamics of Distributed Computation in Complex Systems*, Springer Theses, DOI: 10.1007/978-3-642-32952-4,
© Springer-Verlag Berlin Heidelberg 2013

material), while the fire can also be started in other ways (e.g. overturning a lighted oil stove) [3]. If one never observes the short circuit in the right conditions, without the other fire triggers, the transfer entropy is in fact unable to infer a causal link from the short circuit to the fire.

References

1. T.Q. Tung, T. Ryu, K.H. Lee, D. Lee, Inferring gene regulatory networks from microarray time series data using transfer entropy, ed by P. Kokol, V. Podgorelec, D. Mičetič-Turk, M. Zorman, M. Verlič. in *Proceedings of the Twentieth IEEE International Symposium on Computer-Based Medical Systems (CBMS '07)*, Maribor, Slovenia (IEEE, Los Alamitos, USA, 2007), pp. 383–388.
2. P.F. Verdes, Assessing causality from multivariate time series. Phys. Rev. E **72**(2), 026222–026229 (2005)
3. J.L. Mackie, in Causes and Conditions, ed. by E. Sosa and M. Tooley, *Causation* (Oxford University Press, New York, 1993)

Appendix F
Information Destruction Only Measured in Open Computational Systems

Crucially, the laws "of a *closed* physical system are one-to-one"[4] ([1], p. 78), meaning that in closed physical systems (or the universe as a whole) computational paths[5] do not merge. In other words, there is no irreversible information destruction in closed physical systems. We only measure the *departure* of information from an observed scope in *thermodynamically open* computational systems, where the apparently destroyed information is offloaded (along with energy dissipation) into the external, unobserved environment. The logical computational system and external environment are connected via the physical representation of the computational system (e.g. bit registers): after all, "information is physical" [2]. The information is only destroyed from the scope of the computational system.

Interestingly, this provides an important distinction between the concepts of information modification and information destruction. Were the concepts identical then we would have to accept that there is no information modification in closed physical systems, i.e. that observations of information modification are artifacts of a reduced observational scope in open systems. There is no reason that the concept of this computational operation should be limited as such though.

Furthermore, the fact that we may not be observing the parts of the environment where information is offloaded to should not be taken to mean that we have an impoverished view of the computation itself. While it may ignore the full information output from each variable, it does retain a complete view of the forward computation from the inputs to each variable in the system.[6]

---

[4] Emphasis added only in the inclusion here.

[5] Computational paths in the system's state-space or phase portrait.

[6] This assumes as per footnote 12 on p. 96 that the system is causally closed; i.e. all causal input sources are accounted for within the system—see the breakdown of the information to compute the next state of each variable in Sect. 2.25.

J. T. Lizier, *The Local Information Dynamics of Distributed Computation in Complex Systems*, Springer Theses, DOI: 10.1007/978-3-642-32952-4,
© Springer-Verlag Berlin Heidelberg 2013

References

1. S. Lloyd, *Programming the Universe* (Vintage Books, New York, 2006)
2. R. Landauer, Information is physical. Phys. Today **44**(23), 23–29 (1991)

Appendix G
Circumstantial Evidence of Maximum Coherence in Complex Computation

Certainly, the framework for the information dynamics of distributed computation introduced in Chaps. 3–5 has proven successful in identifying computational structure in local profiles of information storage, transfer and modification. More importantly for our purposes here, we qualitatively observed that those results show the known complex rules, 54 (Fig. 3.4) and 110 (Fig. 3.5), to exhibit the largest amount of *coherent information structure* in their spatiotemporal information dynamics profiles. Specifically, by "coherent information structure" we mean their gliders being coherent information transfer structures and blinkers being coherent information storage entities, since spatiotemporally neighbouring points in these structures have similarly high values of transfer and storage respectively.[7] Rule 18 (Fig. 3.9) contains a smaller amount of less coherent structure in its domain walls. Chaotic rules 22 (Fig. 3.10) and 30 (Fig. 3.11) certainly exhibit all of the elementary functions of computation, but do not appear to contain any coherent structure to their computations. These observations align well with similar explorations of other types of local information structure for these rules, e.g. [1–3].

In this appendix we briefly investigate two fairly obvious possibilities for quantifying coherent information structure in complex systems. The first approach is to check whether the average values of any of the existing measures of information dynamics are suitable for this purpose. Subsequently we examine the spatiotemporal auto-correlation values in profiles of the individual information dynamics. These approaches are shown to be indicators of, or useful circumstantial evidence for the presence of coherent information structure, but do not measure it per se.

G.1 Average Information Dynamics

In CAs, coherent structures have been observed to contain large values of information transfer (for gliders and domain walls) and storage (for blinkers). Similarly, we have observed maximisations of information storage and information transfer near the

[7] See also footnote 3 on p. 80.

J. T. Lizier, *The Local Information Dynamics of Distributed Computation in Complex Systems*, Springer Theses, DOI: 10.1007/978-3-642-32952-4,
© Springer-Verlag Berlin Heidelberg 2013

Table G.1 Table of average information dynamics (all with $k = 16$, and values to 2 decimal places except S^- for rule 110), for several ECA rules

Rule	H	H_μ	A	$T_{j=1}$	$T_{j=-1}$	$T^c_{j=1}$	$T^c_{j=-1}$	S	S^+	S^-	$p(s < 0)$
110	0.99	0.18	0.81	**0.07**	**0.11**	**0.07**	**0.11**	0.98	0.98	**−0.002**	**0.003**
54	1.00	0.27	0.73	**0.08**	**0.08**	**0.19**	**0.19**	0.89	0.90	**−0.01**	**0.03**
22	0.93	0.75	0.19	0.19	0.19	0.56	0.56	0.56	0.62	−0.05	0.09
18	0.82	0.53	0.29	0.01	0.01	0.52	0.52	0.32	0.46	−0.14	0.25
30	1.00	0.99	0.01	0.73	0.01	0.98	0.26	0.75	0.82	−0.07	0.08

Units are in bits (except for the probability $p(s < 0) = p(s(i, n) < 0)$, the proportion of space-time points with negative local separable information). **Bold** font used to highlight values referred to in the main text

critical point in RBNs. We consider here whether the reverse is true: do large values of the information dynamics (or other relationships between them) imply the existence of coherent structures and complex computation? This may be the case in certain phase transitions in a single parameter (such as in large-sized RBNs). However, we are interested in whether such implications hold in general systems, e.g. CAs where there is no known continuous phase transition in a single parameter. The averages for the CAs analysed locally previously were calculated and are presented here in Table G.1.

Certainly large values for the active information storage $A(k = 16)$ are observed in Table G.1 for the known complex rules, and we know that this measure was maximised near the critical state in RBNs. Yet we have seen in Sect. 3.3.2 that the major component of the information storage values in rules 54 and 110 come from their background domains, and it is straightforward to conceive simple oscillating systems maximising this measure without any interacting coherent structures. As such, large information storage remains an indicator of, but not a measure for coherent structure.

Similarly, the apparent transfer entropy was maximised near the critical state in RBNs. In CAs however, we observe that the chaotic rules 22 and 30 have significantly larger values of $T(j, k = 16)$ than the complex rules, so here also the apparent transfer entropy on its own does not measure coherent structure.

That being said, a striking feature of the known complex rules is that the apparent transfer entropy in *each* channel j is a large proportion of the complete transfer entropy $T^c(j, k = 16)$ for that channel. This is true also near the phase transition in RBNs in Sect. 6.1.2. Apparent transfer entropy can only be high where the source has a clear coherent influence on the destination, while complete transfer entropy can separately be high due to *interaction-based transfer*. In the complex CAs, single sources can often influence destinations without needing to interact with other sources, supporting the movement of coherent particle structures and propagation of coherent effects over long distances. Importantly, this occurs for multiple channels, meaning that we have bidirectional travelling coherent structures that should *interact* at some point.

A similar feature is that their separable information $S(k = 16)$ approaches the entropy (again indicating dominance of single-source effects), along with a very low

Table G.2 Autocorrelation $C_{tt'}$ between local transfer entropy values in ECAs separated by 1 step in time and space ($t = t(i, j = 1, n, k = 16)$, $t' = t(i + 1, j = 1, n + 1, k = 16)$)

ECA rule	$C_{tt'}$
110	0.19
54	0.45
22	0.09
18	0.44
30	0.03

proportion of non-trivial information modification events (indicated by an almost vanishing S^- and a small proportion of points $p(s(i, n) < 0)$ with $s(i, n) < 0$). This was also observed to be the case around the critical phase in RBNs in Sect. 6.1.2. Given our knowledge of the importance of these events to computation, their shortage in complex computation initially seems counter-intuitive. However, we suggest that the power of these events lies in their subtlety: used sparingly (or perhaps judiciously in a guided system) they can have high impact in a complex *coherent* computation. Occurring too often they disturb the coherence of the computation which then becomes chaotic; and their impact diminishes.

Similarly, we note that chaotic rules exhibit higher values of the complete transfer entropy itself, in alignment with the complete transfer entropy increasing in the chaotic regime in RBNs. This provides another indication of significant interaction between components eroding the coherent computation in this regime.

Importantly though, these observations quantify neither coherence nor complexity of computation, and again one can conceive of simple systems which produce similar indicators without having coherent interacting structures. That being said, they do provide useful heuristics or *circumstantial evidence* for identifying those properties.

G.2 Autocorrelations in Local Information Dynamics Profiles

Our interpretation of coherence as meaning a logical spatiotemporal relationship between local values suggests that it may be measured via the autocorrelation within profiles of each of their local information dynamics. Table G.2 shows for example the autocorrelation $C_{tt'}$ for local transfer entropy values $t(i, j = 1, n, k = 16)$ separated by 1 step in time and 1 step to the right in space. The separation for the autocorrelation here is the same as the interval across which the local transfer entropy is measured. Notice that the rules exhibiting particles (110, 54 and 18) display the highest correlation values here, since coherent particles exhibit spatiotemporal correlation in the direction of particle motion. Similar results are observed for $t(i, j = -1, n, k = 16)$ with autocorrelation over 1 step in time and 1 step to the left in space.

Again, this observation is a useful heuristic, and parallels the above observations regarding the high proportions of apparent transfer entropy values. However, it still treats the information dynamics independently which misses coherent structure in their interaction. It is also a linear measure, which could miss more subtle types of relationships here. Finally, it is difficult to extend beyond lattice systems (say to

RBNs) where agents are heterogeneous and there are no generic spatial relationships such as “1 step to the right”.

References

1. C.R. Shalizi, R. Haslinger, J.-B. Rouquier, K.L. Klinkner, C. Moore, Automatic filters for the detection of coherent structure in spatiotemporal systems. Phys. Review E **73**(3), 036104 (2006)
2. T. Helvik, K. Lindgren, M.G. Nordahl, Local information in one-dimensional cellular automata, ed by P.M. Sloot, B. Chopard, A.G. Hoekstra. in *Proceedings of the International Conference on Cellular Automata for Research and Industry, Amsterdam*, ser. Lecture Notes in Computer Science, vol. 3305. (Springer, Berlin, 2004), pp. 121–130.
3. J.E. Hanson, J.P. Crutchfield, The attractor-basin portait of a cellular automaton. J. Stat. Phys. **66**, 1415–1462 (1992)

Author Biography

Dr. Joseph Lizier is currently a Postdoctoral Fellow at the CSIRO ICT Centre in Sydney, Australia, in Dr. Mikhail Prokopenko's Adaptive Systems team. Prior to his current role at CSIRO, Joseph was a Postdoctoral Researcher at the Max Planck Institute for Mathematics in the Sciences in Leipzig, Germany (2010–2012), in Prof. Jürgen Jost's Dynamical Systems and Network Analysis, and Cognition and Neurosciences groups. He has also spent 10 years in the telecommunications industry, as a Senior Research Engineer at Telstra Research Laboratories (2001–2006) and Seeker Wireless (2006–2010) in Sydney.

This book is the result of Joseph's PhD thesis in Computer Science. Joseph's PhD degree was awarded in 2010 from The University of Sydney (co-supervised at the CSIRO ICT Centre). This thesis received an Honourable Mention in the CORE Doctoral Dissertation awards for most outstanding Computer Science thesis in Australia, and Joseph was a finalist for the Rita and John Cornforth Medal for an outstanding PhD graduate at The University of Sydney. Joseph has also received a Bachelor of Engineering in Electrical Engineering (with Honours and the University Medal) in 2001 and a Bachelor of Science in 1999 from The University of Sydney.

Joseph has published over 30 journal articles and peer-reviewed conference papers, given invited seminars at several workshops and institutions such as The University of Cambridge, and received the Best Paper award at the IEEE Symposium on Artificial Life in 2011. His research interests include information dynamics

J. T. Lizier, *The Local Information Dynamics of Distributed Computation in Complex Systems*, Springer Theses, DOI: 10.1007/978-3-642-32952-4,
© Springer-Verlag Berlin Heidelberg 2013

in distributed computation, complex systems including cellular automata and random Boolean networks, complex networks and computational neuroscience. His current research focusses on how the properties of information dynamics (presented in this thesis) relate to the complex network structure that underlies the system.

Joseph's full CV and list of publications are available at http://lizier.me/joseph/.

Index

J. T. Lizier, *The Local Information Dynamics of Distributed Computation in Complex Systems*, Springer Theses, DOI: 10.1007/978-3-642-32952-4,
© Springer-Verlag Berlin Heidelberg 2013

S

T

U

V

X

MIX
Papier aus verantwortungsvollen Quellen
Paper from responsible sources
FSC® C105338

If you have any concerns about our products,
you can contact us on
ProductSafety@springernature.com

In case Publisher is established outside the EU,
the EU authorized representative is:
Springer Nature Customer Service Center GmbH
Europaplatz 3, 69115 Heidelberg, Germany

Printed by Libri Plureos GmbH
in Hamburg, Germany